Quantum Physics Meets the Philosophy of Mind

Philosophische Analyse / Philosophical Analysis

Herausgegeben von/Edited by
Herbert Hochberg, Rafael Hüntelmann,
Christian Kanzian, Richard Schantz, Erwin Tegtmeier

Volume / Band 56

Quantum Physics Meets the Philosophy of Mind

New Essays on the Mind-Body Relation in Quantum-Theoretical Perspective

Edited by
Antonella Corradini, Uwe Meixner

DE GRUYTER

ISBN 978-3-11-055473-1
e-ISBN 978-3-11-035106-4
ISSN 2198-2066

Library of Congress Cataloging-in-Publication Data
A CIP catalog record for this book has been applied for at the Library of Congress.

Bibliographic information published by the Deutsche Nationalbibliothek
The Deutsche Nationalbibliothek lists this publication in the Deutsche Nationalbibliografie;
detailed bibliographic data are available on the Internet at http://dnb.dnb.de.

© 2017 Walter de Gruyter GmbH, Berlin/Boston
This volume is text- and page-identical with the hardback published in 2014.
Printing: CPI books GmbH, Leck

♾ Printed on acid-free paper
Printed in Germany

www.degruyter.com

To the memory of our friend E. Jonathan Lowe, who should have taken part in this project

Table of Contents

Preface —— 1

Part I Quantum Physics and the Mind

Henry P. Stapp
Quantum Physics and Philosophy of Mind —— 5

Uwe Meixner
Of Quantum Physics and DOMINDARs —— 17

Godehard Brüntrup
Quantum Mechanics and Intentionality —— 35

Antonella Corradini
Quantum Physics and the Fundamentality of the Mental —— 51

Jeff Barrett
Quantum Mechanics and Dualism —— 65

Part II Quantum Physics, Consciousness, Agency, and Free Will

Ulrich Morhoff
Consciousness in the Quantum World: An Indian Perspective —— 85

Stuart Hameroff
Consciousness, Free Will and Quantum Brain Biology – The "Orch OR" Theory —— 99

Massimo Pauri
Physics, Free Will, and Temporality in the Open World —— 135

Robert Kane
Quantum Physics, Action and Free Will: How Might Free Will Be Possible in a Quantum Universe? —— 163

Peter Jedlicka
Quantum Stochasticity and (the End of) Neurodeterminism —— 183

Name Index —— 199

Subject Index —— 203

The Contributors —— 209

Preface

Looking back at the history of the philosophy of science in the twentieth century, we see that physics was for a long time the scientific discipline that served as a methodological model for all the others. The predominance of physics in the philosophy of science is the main reason for the rise of a philosophical view that goes under the name of "physicalism". This view – old materialism strengthened and rejuvenated by philosophy of science – has imposed, and still imposes, constraints on eligible conceptions in the philosophy of mind.

The question that first caught the attention of the editors of this book was the question of why most philosophical debates on physicalism refer to classical physics, and not to the field that is currently considered to be part of the very foundations of physics, that is to say, to quantum theory. Even though classical physics has been refuted and quantum physics confirmed to a high degree, philosophers rarely ask themselves what consequences can be drawn from a theory that is as solidly anchored in the empirical world as quantum theory happens to be. In striking contrast, the dramatic developments of the neurosciences have caught a large amount of philosophical attention.

Whatever may be the answer to the mentioned question (the myopic concentration on classical physics seems a serious philosophical mistake), the editors of this book believe that the recent developments in the neurosciences have had philosophical reverberations that, as far as a non-reductive philosophical view of the mental is concerned, must be deemed largely negative. Neurophilosophers often argue that the place of mind in nature is occupied by items *other* than mental phenomena, namely, by neurophysiological events, which are thought to be ultimately amenable to *classical physics*. Faced with this situation, the editors ask: What would happen if one took quantum physics seriously? What would happen if one supplemented the philosophical reflections on classical physics and the neurosciences with those on quantum physics? Would it perhaps be possible to give back to the mind its rightful place in nature?

Of course, no scientific theory – and quantum theory is no exception – can claim to be able to confirm or refute, by itself alone, theses and theories that are of a philosophical nature. The editors are well aware that quantum theory itself needs a philosophical interpretation – and that several different philosophical interpretations have been put forward since quantum physics began its course. In spite of this caveat, the editors dare formulate the hypothesis, to be examined in this book, that quantum physics can significantly contribute to solving the puzzle of the mind-body nexus in an essentially dualistic manner.

Ten theorists – philosophers interested in physics, and physicists interested in philosophy – have contributed to this volume on the prospects of a non-reductionist, non-materialist philosophy of mind, which seem to open up once the empirical lessons from quantum theory are non-dismissively taken into account. Five of the essays are more of a general nature, five of them are more specialized; all of the essays are relevant to the key issues: consciousness, action, free will. Readers will discover that the authors' verdicts are as far from blind enthusiasm as they are from biased skepticism.

The event that prompted this book is an international conference at the Catholic University of Milan, Italy, that took place in June 2013. All of the authors of this book were speakers at that conference. The conference was funded by the Thyssen Foundation via the administration of the University of Augsburg, Germany. We would like to thank the Foundation for its financial help, which was also extended to the preparation of this book for publication.

Milan – Augsburg, June 2014

Antonella Corradini
Uwe Meixner

Part I: **Quantum Physics and the Mind**

Henry P. Stapp
Quantum Physics and Philosophy of Mind

1 Introduction

The question before us is whether quantum mechanics can help solve the problems of philosophy of mind.

I believe it can, and my talk will explain how.

What are these problems? They are tied to the ideas of classical physics that prevailed in science during the eighteenth and nineteenth centuries, and were eloquently described by the great 19th century physicist John Tyndall (1874):

> We can trace the development of a nervous system and correlate it with the parallel phenomena of sensation and thought. We see with undoubting certainty that they go hand in hand. But we try to soar in a vacuum the moment we seek to comprehend the connection between them [...] Man as object is separated by an impassable gulf from man as subject. There is no motor energy in intellect to carry it without logical rupture from one to the other.

If there is, indeed, such an impassable gulf, then a primary question is: On which side do we lie?

The belief of most contemporary neuroscientists, and philosophers of mind, is that we lie on the physical side: and that our conscious experiences must therefore be built out of the material stuff of our bodies, and, more specifically, of our brains or nervous systems.

That conclusion draws its scientific support from the principles of classical mechanics, which claimed that the behavior of our bodies could be completely explained without ever mentioning or considering our conscious thoughts, ideas, and feelings.

Huge efforts have been made to understand, rationally, how "man as subject" can arise from the material stuff of classical mechanics—how something like the motions of bouncing billiard balls could *be*, or *produce*, a conscious thought. But many scientists and philosophers now agree that no progress at all has been made in resolving that classical-physics-based mystery.

Sir Karl Popper described the current mainline view in neuroscience as "promissory materialism": with the "promise" being that dogged adherence to the principles of classical mechanics will eventually lead to an understanding of consciousness.

Classical mechanics was, however, found during the twentieth century to be incompatible with a growing host of empirical findings, and was replaced at the fundamental level by quantum mechanics. A key innovation of the new theory was to bring our conscious thoughts into the theory as logically essential parts of the basic underlying dynamics. This quantum approach leads, via the "orthodox" formulation of John von Neumann (1955),[1] to a clean ontological separation between our mental and physical aspects, which, however, become tied together by a dynamical connection. John Tyndall's nineteenth century "impassable gulf" has thus been bridged, during the twentieth century, by replacing an empirically invalid classical physics by empirically valid quantum physics!

But how did such a radical change in the foundations of physics come about?

2 The original "Copenhagen" version of quantum mechanics

Early in the twentieth century a series of theoretical and experimental findings showed that the classical principles that work so well for large astronomical and terrestrial objects, fail to work for their atomic constituents! A new set of laws was found to hold for the atoms. But if we try to apply these atomic laws to the atomic constituents of us human observers, then we usually find that what we experience is altogether different from what the atomic theory predicts!

Specifically, the atomic laws generally entail that the brain of an observer will naturally evolve into a mixture of many different quantum components, each of which corresponds to a different perceptual experience. Yet only one of these perceptions occurs in any actual empirical instance. Consequently, the atomic theory, understood in the ordinary traditional way, fails to agree with experience.

The founders of quantum mechanics resolved this conflict between the atomic laws and human experience by abandoning the conceptual framework that Isaac Newton had created in the seventeenth century. That "classical" way of thinking had, for more than two centuries, been accepted by scientists as the proper foundation of science. But that approach excluded, as a matter of basic principle, any irreducible effect of our conscious thoughts on the behavior of the physically described aspects of the universe.

---

[1] Especially Chapter VI, *The Measuring Process*.

Orthodox quantum theory revokes that exclusion! It converts our conscious experiences from passive spectators to active participants in the creation of our future experiences.

The rational foundation of the new approach was the demand by the creators of quantum mechanics that science be anchored in what we know. But everything we know resides in our experiences. Hence the founders backed away from the idea that the aim of science is to comprehend the reality that *lies behind* our experiences. They focused instead on the structure of our experience itself.

In the words of Niels Bohr:

> In our description of nature the purpose is not to disclose the real essence of phenomena but only to track down as far as possible relations between the multifold aspects of our experience (1934, 18).

Quantum theory was, therefore, originally offered not as theory of "reality", as defined in some abstract classical sense. It was presented, rather, as a practical tool for making predictions about our future experiences on the basis of information derived from our past experiences. Thus human experiences became the basic realities of the theory: the basic realities were shifted from the objective to the subjective side of Tyndall's "impassable gulf".

In this pragmatic approach, we observers are—in order to make the theory useful to us—represented within the theory in the way that we intuitively conceived ourselves, namely as psycho-physical agents that can form intentions based on our own reasons, emotions, and values, and can then physically act to implement those intentions. Von Neumann's orthodox formulation of the theory integrates these features into a rationally coherent understanding of nature itself, and our place within it.

In quantum theory our mental intentions are "freely chosen", in the sense that they are not determined within the theory by prior physical properties. Thus these intentions are allowed by the theory, in its orthodox realistic version, to depend irreducibly on consciously felt values that are described in psychological rather than in physical terms! That means that quantum theory violates a core idea of classical mechanics: it allows our mental "free choices" to influence our physical actions, yet not be fully determined by prior physically describable properties. Thus the philosophy-of-mind concept of "physicalism" fails: the demand that "all is physical" is not only not entailed be contemporary basic physics; it is also strictly incompatible with the orthodox realistic formulation of it.

The general logical form of the empirically validated quantum mechanical dynamics is this. Before each perception, the observer must choose and perform a probing action. That action effectively asks Nature whether or not the system being observed has a specified physical property. The existence of this physical property will be communicated to the observer by a specified-by-the-observer response from nature. This possible positive ("Yes") response is chosen by the observer, not by Nature. If Nature's answer is "Yes", then two things immediately happen: the observer will experience the observer-selected response, and the system that is being observed will immediately acquire the specified physical property. If Nature's answer is "No", then the observer will experience nothing, in connection with this negative answer, but the physical possibilities will be reduced by the exclusion of the "Yes" possibility.

Nature's choice between the two possible responses, "Yes" or "No" is asserted to conform to a certain quantum statistical rule. But the observer's choices are, in both the pragmatic and orthodox realistic versions, supposed to arise from the observer's motives and values! Thus the conception of "the user" conforms, in both versions, to the user's life-experience-based idea of himself or herself.

Whereas classical physics renders life meaningless, by asserting that we are, effectively, mindless mechanical puppets, acting out a pre-choreographed script, quantum mechanics restores meaning by allowing, and indeed causing, one's own experienced future to be directly influenced by one's own value-based consciously felt efforts!

A key feature of this quantum observation process is that the property chosen by the observer is something that the observed system possesses *after* the process is completed, but may not have possessed *before* the process was initiated.

For example, the quantum state of an observed system before the probing action might correspond to a "wave" that is spread out over a large spatial region, whereas, after the response, the state might correspond to the system's being confined to a particular atom-sized region. Such a "collapse of the quantum state" provides an immediate resolution of the wave-particle duality problem.

We see here the beginnings of the quantum bridge over Tyndall's "impassable gulf" between "man as subject" and "man as object". For the observer's conscious choice, which lies on the subjective side of the gulf is causally affecting the objective physically described world, which lies on the other side.

3 John von Neumann's orthodox version of quantum mechanics

The founders dodged various puzzling metaphysical issues by claiming to be providing merely a practical tool that works in practice. But philosophy of mind cannot evade basic metaphysical questions.

The eminent mathematician John von Neumann faced the difficulties head-on, by converting the original Copenhagen pragmatic version of quantum mechanics into a form that can be regarded as an empirically validated putative theory of an interactive psycho-physical reality.

But what changes did he institute?

The original "Copenhagen" way of describing the collapse process was tied to a mysterious thing called the "Heisenberg cut". Everything lying "below" this cut was supposed to be described in the mathematical language of quantum mechanics, whereas everything lying "above" the cut was described either in the language of classical physics, or in psychological or mental terms. The idea was that a *practical* account must accommodate our possible mental intentions and free choices, and also our descriptions of—in Bohr's words—"what we have done and what we have learnt" (1934, 3). Those things were described in mental and classical terms, whereas their atomic underpinnings were described in terms of the quantum mathematics.

This Heisenberg cut was "movable": its placement depended on what practical use was to be made of the theory. But that "movability" meant that the same physical thing could be described in two logically incompatible ways—either classically or quantum mechanically—depending on the practical application.

Such an inconsistency might be all right for a purely practical theory, but it is not acceptable for a putative description of reality itself.

A principal move made by von Neumann was to show that the Heisenberg cut could be moved all the way up, so that reality was unambiguously separated into a purely mental part, and a part described in terms of the mathematical language of quantum mechanics. The external measuring devices became parts of that latter world, while the "classical descriptions" of these devices became unambiguously identified as aspects of the perceptions of observers.

The mentally described experiences were kept fixed, while the Heisenberg cut was shifted up, step-by-step, until all atomically constituted things, including our physical bodies and brains, lie below the cut, and hence are described in the mathematical language of quantum mechanics. The observer's mental aspects are *preserved* during this shift of the cut, and they are eventually pushed completely out of the physically described universe: mind-matter separation is achieved within a theory that dynamically encompasses both.

These preserved mental aspects were called abstract "egos" by von Neumann. They are mental in character, and are ontologically separated from the physical world. Yet each such ego retains a quantum dynamical linkage to an associated physical brain. Thus Tyndall's "impassable gulf" between "man as subject" and "man as object" has been bridged by rigorous quantum mathematics: Von Neumann converted what had originally been offered as a mere practical tool into a rationally coherent putative description of a dynamically integrated psycho-physical reality!

Von Neumann's formulation eliminates the notion that mere "bigness" can somehow cause a collapse. After all, how big is big? Von Neumann's formulation ties the collapse not to something as nebulous as "big", but to something that, according to the theory, is separate from the physical world—namely consciousness! And his theory specifies the place where consciousness acts—namely on the brain of the observer.

Our egos are thus ontologically separated from our bodies. But then how can they control our bodily movements?

4 The causal effectiveness of our minds

It might seem that a mere capacity to pose questions and register answers would leave our mental egos just as helpless and impotent as before. But the quantum mechanical process of posing questions and receiving responses is not like the classical mechanical process, in which our observations have no physical effects. In QM, the observer's free choices of which question to ask play a critical role in determining which potential atom-based property will become actualized.

In QM, the observer asks Nature a yes/no question about the state of a system. If Nature's response is "Yes", then *after* this response is delivered the system will definitely have the property *that the observer freely picked*.

Normally, this dependence of the post-observation properties of the system that is being probed upon the observer's choice of question does not give the observer any effective control over the observed system. That is because Nature's response can be "No".

However, there is an important situation in which, *according to the quantum rules*, the "No" answers will be strongly suppressed. In that case, the free choices made by the observer *can* exert effective control over the system being probed—which, in von Neumann's theory, is the brain of the observer.

Suppression of the "No" responses is predicted if an initial "Yes" response is followed by a sufficiently rapid sequence of posings of the same question. In that case the observer becomes empowered, by his own free choices, to hold stably in place a chosen brain activity that normally would quickly fade away.

This effect is the celebrated "Quantum Zeno Effect", which was linked by Sudarshan and Misra to the paradox of the arrow in flight posed by the Greek philosopher Zeno of Elea: at each instant the arrow is at rest, so how can it move?

This "paradox" is no longer puzzling to scientists. But the quantum dynamics brings it back into play, by claiming that if the quantum observations—which affect the system being observed—become increasingly rapid, then the motion of the system being observed becomes increasingly slowed. In the limit of infinitely rapid observations the observed system becomes frozen in place, like Zeno's arrow.

The important dynamical change in the role of us observers was repeatedly emphasized by Bohr and the other founders, in statements such as:

> in the drama of existence we are ourselves both actors and spectators (1949, 236).

This change in our role in the unfolding of reality vindicates William James's commitment to rationality:

> It is to my mind utterly inconceivable that consciousness should have nothing to do with a business which it so faithfully attends (1950, 136).

5 The essential point

The purpose of the above account of the twentieth-century replacement of classical physics by quantum physics is to make it clear that the foundation within science for both the philosophy-of-mind concept of "physicalism" and the neuroscience idea of "promissory materialism" has now evaporated: Quantum physics constitutes a refutation of the idea that the motions of the atoms in our brains are, like the motions of planets in our solar system, independent of our mental intentions. Quantum physics elevates our conscious intentions from physically impotent side effects, of a physically predetermined evolutionary process, to dynamically essential instigators of our physical actions! The intensive twentieth century efforts by some influential scientists and philosophers to curtail the rational advance of basic science by clinging to simple nineteenth century ways of understanding nature has borne no fruit. It is time, now, to accept the quantum advance in our understanding of nature not only in technology but also in neuroscience and philosophy of mind.

6 Impact on philosophy of mind

Quantum mechanics contains, in addition to a mathematically described physical part, also our psychologically described mental parts, with these two parts rationally tied together by a specified dynamical process. That achievement is precisely what philosophy of mind, and philosophy in general, has been seeking ever since their inceptions: a rationally coherent understanding of a reality that is inclusive—that rationally encompasses simultaneously both the mental and physical parts of nature, and that, building on a science-based empirical foundation, explains how a person's mental intentions and physical behaviors are related to each other!

Some physicists have tried to eliminate the causal effects of the mental, and thereby to revert to the seventeenth century classical ideal. That endeavor may have some practical utility in the pragmatic treatment of purely physical experiments involving external measuring devices, with no effort to account for "our knowledge" of what is physically happening. But from a deep philosophical perspective those efforts are retrograde because they leave out of the ontology the only thing that we really know exists, namely our conscious experiences. The deeper philosophically important endeavor is not to *exclude* the known-to-exist mental part of reality from science. It is rather to bring mental realities into science in a rationally coherent way! That deeper endeavor is an important part of what von Neumann's orthodox theory is all about!

7 Appearances are deceiving

The question naturally arises why most neuroscientists and philosophers interested in the mind-brain connection choose to ignore the very pertinent replacement of classical physics by quantum physics.

One reason, of course, is the power of inertia and authority. Another is the unfamiliar mathematics and logic. More important is the fact that physics textbooks follow the pragmatic Copenhagen tack, in which the quantum collapses are imagined to occur at external measuring devices, rather than in our brains. Physicists, in general, prefer not to mix their physics with psychology, and hence to push aside, as much as possible, the complications associated with our knowledge, and hence to remain content with a purely pragmatic stance.

But probably the most important inhibitor of the embrace of quantum mechanics is the fact that the orthodox theory entails that the seeming validity of classical ideas at the level of visible-sized properties is illusory: According to orthodox quantum mechanics "Appearances are deceiving!" They are profoundly deceiving!

In orthodox quantum theory, the world of tables and chairs and other atomically constituted parts is considered to be fully quantum mechanical. But that means that, in spite of its classical material appearance, the *macroscopic* physical world is "really" bundle of potentialities pertaining to *what will appear to us observers if someone actually looks. Perceivable properties* become actual only insofar as actual perceptions actualize them.

The normally observed "classical" appearance of the visible world is, according to the orthodox theory, created by all of the observations that have been made over the course of the history of the universe. Those conditions are very restrictive. But they still allow a lot of quantum mechanical uncertainty for the status of perceivable-sized things that are not actually perceived.

Our brains, for example, are highly quantum mechanical. Large amounts of quantum uncertainty are introduced by the passages of ions through ion channels. The small spatial diameters of these channels entail large uncertainties in the velocities of the ions emerging from them. A living person's brain is therefore a generator of huge amounts of quantum uncertainty. This uncertainty can percolate up to the macroscopic level without being perceived either by the person himself or by anyone else. *Brains must therefore be treated quantum mechanically*: That is what permits the behavior of a person's brain to be significantly influenced by the free choices made by that person's own conscious mind.

Weird as this quantum feature might seem to scientists steeped in Newtonian physics, it is where quantum mechanics rationally leads. It is in complete accord with all human experience, including our experience-based understanding of ourselves. And it is in line with a certain idea of parsimony that would not allow Nature to encumber itself with a highly developed conscious aspect that can make no difference in what actually happens.

8 Nonlocality and the nonmaterial character of nature

In spite of the accuracy of the quantum predictions pertaining to our normal everyday experiences with ordinary objects, it is not easy to accept the idea that perceivable things that are unperceived differ greatly from actually perceived things... Why should anyone believe such an anti-intuitive idea?

A fundamental feature of quantum mechanics is its inescapable need for faster-than-light transfer of information. Einstein believed this feature to be just a peculiar property of the statistical mathematical formalism; not a property of a putative yet-to-be-discovered underlying reality. However, it has been rigorously proved, by arguments *that make no reference at all to any microscopic property*, that quantum mechanics is incompatible with the general assumption that information about choices made by human agents in various experimental regions cannot be transmitted faster than light to distant experimental regions (Stapp 1977, 191–205; 1979, 1–25; 2010, 198–200).

These proofs rule out the validity of a conception of reality that conforms, at the macroscopic level, to the classical principles of special relativity theory.

9 Retro-causal determination of records of the past

Quite apart from logical proofs of the failure of the classical-physics-based notion of materialism, there are also directly observable phenomena, involving the "appearances" of backward-in-time actions. The quantum collapses have a certain sort of "effective" retro-causal action. The "quantum collapses" not only pick out what actually happens from a set of potentialities for what might happen. They also eliminate from the *records* of the past all traces of properties that led to the possibilities that were eliminated by Nature's choice. Thus the surviving records of the physical processes leading up to the collapse event exhibit only those parts of the past that lead up to what actually did happen: the other parts of the stored records of what was happening prior to the collapse disappear without a trace. As Stephen Hawking and Leonard Mlodinow succinctly put it in their recent book *The Grand Design*:

> We create history by our observations, rather than history creating us (140).

A large number of experiments have revealed the existence of various retro-actions directly *at the macro-level of perceivable-sized effects*. One kind of example consists of a change in the size pupil of the eyes of human subjects slightly before a random-number-generator-timed shocking stimulus is applied! Another kind of example is a sudden increase in skin-conductance slightly *before* a shocking visual stimulus is shown to a human subject.

These retro-effects are incompatible with a material world governed by the principles of classical physics. The precepts of "Promissory Materialism" are thus —*directly at the level of visible phenomena*, and without reference to quantum theory—irreconcilable with apparently mounting scientific evidence.

But the retro-causal effects can be consequences, within a strictly forward-in-time evolving universe, of selections of *which* records of what was happening in the past *survive* the collapse. This collapse creates not only the reality existing at the present global instant "now", but also, via the Schroedinger equation, a new set of potentialities for the future, and also a revised record of the past.

The point, restated, is that, according to orthodox quantum mechanics, the evolving reality is created by a strictly forward-in-time indeterministic process that produces records of the past that exclude the records of the processes that led to the possibilities that were eliminated by the collapse.

10 Conclusion

The failure of classical mechanics at the level of the atoms led to its replacement by quantum mechanics. The basic change wrought by that move was to separate our minds ontologically from our brains, and to convert our minds from puppets controlled by our brains to bona fide players in the game of "Creation of our joint experienced future".

We in the West live in a society that rests heavily on the idea of our own human nature that was fabricated by classical mechanics. Our teachers teach it; our pundits proclaim it; our courts uphold it; our institutions and governmental agencies base their decisions upon it. And we ourselves can be disheartened and inhibited by the meaninglessness of our lives that this incessant message implies. That pernicious fable falsely attributes to science the fiction that we cannot by our mind-driven actions create a better world for ourselves and our progeny.

References

Bohr, N. 1934, *Atomic Theory and the Description of Nature*, Cambridge: Cambridge University Press.

Bohr, N. 1949, "Discussions with Einstein on Epistemological Problems in Atomic Physics", in P.A. Schilpp (ed.), *Albert Einstein Philosopher-Scientist*, New York: Tudor Publishing Company.

Hawking, S. and Mlodinow, L. 2010, *The Grand Design*, New York: Bantam.

James, W. 1950, *Principles of Psychology*, New York: Holt 1890, reprinted New York: Dover.

von Neumann, J. 1955, *Mathematische Grundlagen der Quantenmechanik*, Berlin: Springer 1932, translated by R.T. Beyer, *Mathematical Foundations of Quantum Mechanics*, Princeton: Princeton University Press.

Stapp, H.P. 1977, "Are Superluminal Connections Necessary?", *Il Nuovo Cimento* 40B: 191–205. Explained in greater detail in: http://www-physics.lbl.gov/~stapp/Appendix-1.doc

Stapp, H.P. 2010, Mindful Universe, 2nd Ed., Berlin: Springer.

Stapp, H.P. 1979, "Whiteheadian Approach to Quantum Theory and the Generalized Bell's" Theorem', *Foundations of Physics* 9: 1–25. Available at: www-physics.lbl.gov/~stapp/Whitehead-Bell-1979.pdf

Tyndall, J. 1874, "The Belfast Address", *Nature* 10: 309–319.

Uwe Meixner
Of Quantum Physics and DOMINDARs

"DOMINDAR" is an acronym for "Detector of Macroscopic Indetermination, and Restrictor". In my paper "New Perspectives for a Dualistic Conception of Mental Causation",[1] I presented the hypothesis that the brain, taken together with the entire nervous system, is a DOMINDAR. I argued for this hypothesis by inference to the best explanation: The fact that there are brains and that they obviously are a widespread survival asset in the animal kingdom, produced and perfected by evolution in the course of millions of years, *is best explained* by the hypothesis that (*a*) there is macroscopic indetermination in the physical world that is relevant for the survival and well-being of animals, and that (*b*) their brains serve to detect this indetermination (making essential use of the sensory system) and to restrict it (making essential use of the motor system) in ways that are advantageous for the animals' survival and well-being. If brains are DOMINDARs, it is a further question whether they are DOMINDARs *in their own right* or, on the contrary, *instrumentally*. The latter alternative, if adopted, leads to the further hypothesis of *natural souls*—of souls that are not supernatural beings but a part of nature, each soul serving as an at least minimally rational decision maker for an animal, each doing so by using the DOMINDAR which is the animal's brain. I have defended this further hypothesis in several of my publications.[2] The basic fact that points in the direction of natural souls is that there are not only brains, likely to be DOMINDARs, but also consciousnesses produced by those brains, consciousnesses that each have a subject of consciousness. It is most likely that consciousnesses and subjects of consciousness are not produced as causally inert epiphenomena. What, then, is their likely causal *function*?

Instead of going into this question (my disquisitions would amount to a defense of naturalistic, evolutionary and interactionist *substance-cum-consciousness dualism*), I intend in this paper to present the conceptual basics of DOMINDARs, and to display the juncture where in DOMINDAR-theory quantum physics enters the scene. I will do so more or less abstractly. For illustration, I will use simple "abstracts" of DOMINDARs. The two *burning* questions of DOMINDAR-theory are of course: (1) What, precisely, does it mean to detect macroscopic indetermination and how, in principle, is this detecting implemented? (2) What, precisely, does it mean to restrict macroscopic indetermi-

---

[1] Meixner 2008.
[2] Meixner 2004, 2006, 2010.

nation and how, in principle, is this restricting implemented? I will delve into these questions—but I will do so more or less abstractly.

At the heart of every DOMINDAR is what I call a "REACTOR".[3] Every device of the type "Do something at one end of the thing and be sure (if all is well with the thing) to get something else at the other end" is a REACTOR (but not only such devices). It is obvious that a considerable part of our everyday life is filled by our manipulation of REACTORs: cars, pianos, dish washers, computers, but also simple tools—like hammers, knives, forks—are REACTORs. A REACTOR can be abstractly represented by a non-empty set of ordered pairs, which is such that the first member of each pair in that set is a possible input-state, *determining*, if it occurs at a time *t* and the REACTOR functions well,[4] the occurrence of a certain output-state at time $t+\delta$; that output-state is the second member of the pair. If pairs in the set differ with respect to their first member, then they also differ with respect to their second. All the pairs in the set do not differ with regard to the reaction time, δ: δ is the same for all of them.[5] The possible input-states extractable from the set (that is, the possible inputs of the REACTOR) are exclusive of each other: they cannot co-occur; and the possible output-states extractable from the set (that is, the possible outputs of the REACTOR) are also exclusive of each other: they cannot co-occur. No output-state of a REACTOR is an input-state, and vice versa (of course). If a REACTOR has a finite number of possible input-states, then they can be named and listed completely: $IP_1, ..., IP_N$, and the corresponding possible output-states can also be named and listed completely (in the order of their correspondence to the already listed inputs): $OP_1, ..., OP_N$. This given, the REACTOR can simply be represented by a finite set of conditional statements (*which are true if the* REACTOR *functions well*), each of which has the following form:

If IP_k occurs at t, then OP_k occurs at $t+\delta$ (the index k is to be taken from 1, ..., N).

A *one-reaction* REACTOR is a REACTOR that can be represented by *one* such conditional. A *two-reactions* REACTOR is a REACTOR that can be represented by *two* such conditionals. A *finitely-many-reactions* REACTOR is a REACTOR that can be

---

[3] "REACTOR" is not an acronym, but the word is nevertheless capitalized since it has a special sense here: the sense needed for describing DOMINDARs.

[4] The well-functioning of a REACTOR requires the right inner *and* outer facts, the right "circumstances" (within and outside the REACTOR).

[5] This is a restriction serving the restricted purposes of this paper. In a general theory of REACTORs one would also have to consider REACTORs with varying reaction times.

represented by N such conditionals (for some natural number N). An *infinitely-many-reactions* REACTOR is a REACTOR that cannot be represented by a finite set of conditionals of the above form. It is an interesting fact of our everyday life that we use infinitely-many-reactions REACTORs (complex ones, like cars, simple ones, like hammers) without giving it a moment's thought. In constructing *such* REACTORs, are we not envisaging infinitely many alternative possibilities in macroscopic physical reality, all open to us at one and the same moment of time, it being up to us which of the possibilities will be *the one* that is going to be *actual* (or *real*)?

Yes, the existence of infinitely-many-reactions REACTORs with macroscopic output-states *points in the direction* of *macro-indeterminism*. Indeed, already the existence of *any* more-than-one-reaction REACTOR with macroscopic output-states does so. But of course the existence of such REACTORs does not by itself *prove* macro-indeterminism. A well-functioning REACTOR is either idle or active. It is idle if none of its possible input-states occurs; it is active if one of its possible input-states occurs. It may be (completely) predetermined at what time the REACTOR is well-functioning and idle, and at what time it is well-functioning and active, and *in what way* it is active when it is well-functioning and active. Thus, the existence of any REACTOR with macroscopic output-states, no matter *what* is their number, is entirely compatible not only with macro-determinism, but also with *complete* determinism.[6]

The simplest REACTORs with more than one possible output are the *two-reactions* REACTORs. Their form of abstract representation is this:

If IP_1 occurs at t, then OP_1 occurs at $t+\delta$.
If IP_2 occurs at t, then OP_2 occurs at $t+\delta$.
(IP_1 and IP_2, OP_1 and OP_2 cannot co-occur.)

Note that the mechanism figuring in the thought-experiment which is known as "Schrödinger's Cat" is a two-reactions REACTOR. It is a two-reactions REACTOR of the following *special* type:

If IP_1 occurs at t, then OP_1 occurs at $t+\delta$.
If non-IP_1 occurs at t, then non-OP_1 occurs at $t+\delta$.

For a REACTOR of this type, the requirement that its input-states cannot co-occur and that its output-states cannot co-occur is automatically fulfilled. Moreover,

---

[6] Consider also, in this context, that any REACTOR with more than one possible reaction can be regarded as a set of two or more *one-reaction* REACTORs.

for a REACTOR of this type, the input-states exhaust the space of possibilities at a given time, and so do the output-states. In addition to these features, the REACTOR figuring in *Schrödinger's Cat*—the SCREACTOR, for short—has further specialties. For one thing, its two input-states are microscopic, whereas its two output-states are macroscopic. The most important specialty of the SCREACTOR is, however, that it is not predetermined which of its two possible inputs will be realized at time t. And this is not just a supposition belonging to the set-up of the thought-experiment *Schrödinger's Cat*; for it is *an accepted scientific fact* that it is not predetermined whether this particular radium atom decays at t (which is one of the two possible inputs of the SCREACTOR), or not (which is the other possible input of the SCREACTOR).

It is provable for each well-functioning finitely-many-reactions REACTOR that if it is not predetermined which of its output-states occurs, that then it also not predetermined which of its input-states occurs.[7] But it is not provable for each well-functioning finitely-many-reactions REACTOR that if it is not predetermined which of its input-states occurs, that then it is also not predetermined which of its output-states occurs. In the special case of the SCREACTOR, however, provided it is well-functioning, it *is* provable: If it is not predetermined which of its input-states occurs, then it is also not predetermined which of its output-states occurs.[8]

---

[7] Suppose the occurrence at t of IP_k of a well-functioning finitely-many-reactions REACTOR is predetermined (and therefore, given the definition of a REACTOR, the *non-occurrence* at t of all other input-states of the REACTOR is also predetermined). Hence the occurrence at $t+\delta$ of OP_k is predetermined (as is the *non-occurrence* at $t+\delta$ of all other output-states of the REACTOR); this is so because of the truth of "If IP_k occurs at t, then OP_k occurs at $t+\delta$", in which "if A, then B" is to be taken in a sense that secures the transfer of predetermination from the *protasis* to the *apodosis*. Therefore: If it is predetermined *which* of the input-states of the REACTOR occurs at t, then it is also predetermined *which* of its output-states occurs at $t+\delta$. Therefore (via contraposition, since "if A, then B" is to be understood in such a sense that contraposition is valid for it): If it is not predetermined which of the REACTOR's output-states occurs at $t+\delta$, then it also not predetermined which of its input-states occurs at t.

[8] Suppose it is not predetermined which of the input-states of the SCREACTOR—supposed to be well-functioning—occurs at t. Suppose, moreover, it is predetermined which of the output-states of the SCREACTOR occurs at $t+\delta$. There are two cases under this latter supposition. In case the occurrence at $t+\delta$ of OP_1 is predetermined (case 1), the non-occurrence at $t+\delta$ of non-OP_1 is predetermined, and therefore—because of the truth of "If non-IP_1 occurs at t, then non-OP_1 occurs at $t+\delta$", employing contraposition—the non-occurrence at t of non-IP_1 is predetermined. But this means that the occurrence at t of IP_1 is predetermined—*contradicting the initial supposition*. In case the occurrence at $t+\delta$ of non-OP_1 is predetermined (case 2), the non-occurrence at $t+\delta$ of OP_1 is predetermined, and therefore—because of the truth of "If IP_1 occurs at t, then OP_1 occurs at $t+\delta$", employing contraposition—the non-occurrence at t of IP_1 is predetermined. But this means that the occurrence at t of non-IP_1 is predetermined—*contradicting the initial supposition*. The

Thus, it is neither predetermined that the cat at the output-end of the well-functioning SCREACTOR is alive shortly after time t nor predetermined that it is not alive shortly after t *if* (and only if) it is neither predetermined that the radium atom at the input-end of the SCREACTOR decays at t nor predetermined that it does not decay at that time. And therefore the existence of the well-functioning SCREACTOR would prove the existence of physical macro-indetermination—if the existence of *micro*-indetermination in radioactivity is accepted (but the existence of micro-indetermination in radioactivity is quite uncontroversial).

The SCREACTOR is a remarkable REACTOR. But it is far from being a DOMINDAR. It is illuminating to consider what would be necessary for making a DOMINDAR out of the SCREACTOR. The cat at the output-end of the SCREACTOR is interested in continuing to live; this we can take for granted. *If* the cat knew, due to additional features of the set-up, that it is neither predetermined that she is alive at $t+\delta$ nor predetermined that she is not alive at $t+\delta$, and also knew how to restrict—in fact, *abolish*—this macro-indetermination by abolishing the correlated micro-indetermination (which is not only entailed by the macro-indetermination but is also the sufficient basis for it, *given* the well-functioning of the SCREACTOR, a well-functioning here presupposed), *then* the whole system would be a DOMINDAR, namely, an *instrumental*—though highly artificial—DOMINDAR *of the cat*. The cat would be—very artificially—in the role of a *natural soul*. If the cat in the system—*if* she had the knowledge just described—made it happen that the radium atom does not decay at t, then she would thereby guarantee her being alive at $t+\delta$; and if she made it happen that the radium atom decays at t, then she would thereby guarantee her being dead at $t+\delta$. It would be up to her whether she is dead or alive at $t+\delta$. But, given her interest in survival, she would of course choose the *first* of the indicated two alternative ways of restricting the macro-indetermination in question *by* restricting the correlated micro-indetermination. Unfortunately, there is no way known to cats, or to humans, of how to determine a radium atom's decay, or non-decay, at a given time. It is, therefore, impossible to make a DOMINDAR out of the SCREACTOR.

In abstract terms: A device X is a well-functioning *two-ways* DOMINDAR if, and only if, the following conditions are fulfilled:

---

conclusion on the basis of the initial supposition must therefore be this: it is (after all) *not* predetermined which of the output-states of the SCREACTOR occurs at $t+\delta$. And therefore we have: If it is not predetermined which of the SCREACTOR's input-states occurs at t, then it is also not predetermined which of its output-states occurs at $t+\delta$.

1. The central part of X is a well-functioning two-reactions REACTOR, represented by the following two (predetermination transferring) conditionals:

   If IP_1 occurs at t, then OP_1 occurs at $t+\delta$.
   If IP_2 occurs at t, then OP_2 occurs at $t+\delta$.
   (IP_1 and IP_2, OP_1 and OP_2 cannot co-occur.)

2. IP_1 and IP_2 are physical micro-states, and OP_1 and OP_2 are physical macro-states.
3. X is equipped with a well-functioning *detector* of occasions on which it is neither predetermined that OP_1 occurs nor predetermined that OP_2 occurs.
4. X is equipped with a well-functioning *determiner* for occasions on which it is neither predetermined that IP_1 occurs nor predetermined that IP_2 occurs.
5. The determiner of X determines either the occurrence at t of IP_1 or the occurrence at t of IP_2, *if* the detector of X detects that neither the occurrence at $t+\delta$ of OP_1 nor the occurrence at $t+\delta$ of OP_2 is predetermined and *if* X acts on the occasion.

Suppose now that X is a well-functioning two-ways DOMINDAR, and suppose that neither the occurrence at $t+\delta$ of OP_1 nor the occurrence at $t+\delta$ of OP_2 is predetermined. For this reason, and because the REACTOR of X is well-functioning, it follows that it is neither predetermined that IP_1 occurs at t nor predetermined that IP_2 occurs at t (see the proof in footnote 7). Suppose X is doing the job it is well-equipped for. Then its detector will detect the OP_1/OP_2 macro-indetermination at $t+\delta$, and its determiner will restrict—in fact, abolish—this OP_1/OP_2 macro-indetermination *by* determining either that IP_1 occurs at t, or that IP_2 occurs at t. For it is guaranteed—given the well-functioning of the REACTOR of X—that if IP_1 occurs at t, then OP_1 occurs at $t+\delta$, and if IP_2 occurs at t, then OP_2 occurs at $t+\delta$.

Several things are important to note here, which are already quite apparent in the simple case of a two-ways DOMINDAR: (i) A DOMINDAR restricts—in the special case: abolishes—(an instance of) *macro*-indetermination *via* restricting (an instance of) *micro*-indetermination. (ii) It does not need to detect this micro-indetermination; rather, it may be the case that the micro-indetermination is not only instrumental in restricting the macro-indetermination but also in detecting it (see below). (iii) The determiner of the DOMINDAR acts, if it acts, at the time at which the actual (but not predetermined) input-state of the REACTOR of the DOMINDAR occurs; the detector of the DOMINDAR acts, if it acts, *not later* than that time. (iv) A DOMINDAR is not per se *rationality-guided*; in order to be rationality-guided, a DOMINDAR must restrict the macro-indetermination it detects in accordance with the *interests* of a being that does have interests in the macroscopic physical world—for example, the interest to continue to exist. This interested entity may, of course, be the DOMINDAR itself; or the interested entity may be an entity of which the DOMINDAR is an organ; or the interested entity may be an entity that knows (but need not be articulate about it) that its own existence depends completely and utterly on the existence of the organism of which the DOMINDAR is an organ—an entity which is not the incarnation but, so to speak, the *empsychization* of the life-interests of the organism.

The three central questions concerning DOMINDARs are these: (A) How does the detector of a DOMINDAR work? (B) How does the determiner (or *restrictor*: see footnote 13) of a DOMINDAR work? (C) Are there any DOMINDARs? The first two questions are utterly difficult to answer. I will make some suggestions, but perhaps questions (A) and (B) are impossible for us to answer. This may give one the idea that there are no DOMINDARs—along the lines of a very familiar, but hardly ever explicitly avowed pattern of anthropocentric thinking, according to which something that *we* just cannot understand must be assumed to be simply non-existent. This pattern of thinking is difficult to defend even in cases that are favorable to its application; in the case of DOMINDARs, however, the pattern is just about indefensible. *Consider*:

In a room with no easy way out, an evil person has left me alone with a bomb that is set to explode in ten seconds, at 12 o'clock. But I can run to the bomb and touch it. The evil person was confident that I do not know how to prevent the bomb from exploding. Fortunately, I happen to know. If I turn this particular little switch on the casing of the bomb to the left, the bomb will inevitably not explode; if I don't, it will inevitably explode. This I know. Not wishing to die, I therefore run to the bomb and turn the switch to the left. The bomb does not explode, and I survive.

Obviously, the bomb was not predetermined to explode, for it did not explode. Was it predetermined *not* to explode? Hardly, for it would inevitably have exploded if I had not intervened at the very last moment, a split second before the blast. But if the bomb was neither predetermined to explode nor not to, then I myself was neither predetermined to turn the switch to the left nor not to (for—given the nature and well-functioning of the mechanism of the bomb—if there had been the latter predetermination, for the one behavioral outcome or the other, then there would certainly have been also the former predetermination, for the one or the other of the corresponding "pyrotechnical" outcomes). If so, did not my brain—either in its own right, or as an instrument of my soul[9]—detect *this latter* macro-indetermination, and did it not restrict—in fact, abolish—it in the way favorable to my survival by abolishing, in the depth of my brain, a micro-indetermination that corresponds to that macro-indetermination in the manner previously described, the descrip-

---

9 I am my soul, in a certain sense of "I". In another sense of "I", I am this entire human being, this unity of body and soul, and in this sense, I am *not* my soul.

tion being in terms of input-output conditionals and their logical consequences? In short, did not my brain act as a well-functioning two-ways DOMINDAR? I, for my part, am very much tempted to concur. The only hitch is that I do not know *how* the brain, or any other candidate for being a DOMINDAR, detects macro-indetermination—this is *the detection-problem*—, and how it restricts, even abolishes, micro-indetermination—this is *the determination-problem*.

The center of the detection-problem is that a state of indetermination—whether *macro* or *micro*—is a state of pure possibility, not a state of actuality. How does the brain detect that there are alternative, hence incompatible, unactualized possibilities relative to the same future moment of time? Only something that is actual can be detected. It seems, therefore, that the detection of indetermination is impossible.

At this point quantum physics comes to the rescue: it suggests a way out of the difficulty. States of indetermination are, in themselves, states of pure possibility; but fortunately for the DOMINDAR-project they are also, so to speak, *incarnated* in certain states of actuality—namely, in states of quantum-physical superposition. A state of indetermination may be detected via detecting the state of quantum-physical superposition which *incarnates* that state of indetermination.

Superposition states are notoriously hard to describe as soon as one moves beyond the mathematical formalism. The cat in Schrödinger's thought-experiment is taken to be in a macroscopic superposition state—and in popular books on quantum physics the reader is told that the cat in that state is dead and alive, and/or neither dead nor alive. Inadequate as such descriptions certainly are,[10] they nevertheless show that states of quantum-physical superposition are fit to *incarnate* states of indetermination. The incompatible, alternative possibilities of a state of indetermination are *superposed* in the *incarnating* superposition state—which is a state of actuality; the *incarnated* state of indetermination, a state of pure alternative possibilities, is "decided on"—is in one way or another replaced by a classical state of actuality—if, and only if, the incarnating superposition state mutates into a *definite* state of actuality: in selection-actualization of precisely one of the alternative (and precise) possibilities of the state of indetermination it incarnates (by superposition of those alternative possibilities).

So far, so good. But there are further difficulties, further aspects of the detection-problem. Would not the very detection of a superposition state automatically make it mutate into definiteness, and make it mutate into definiteness in an uncontrollable way (that is, make it "collapse")? And are there *macroscopic* superposition states? Even if there are, this, by itself, is not enough for the feasi-

---

[10] The best ordinary-language description of superposition states is this: they are *actual, but ontologically vague* states.

bility of the central part of the DOMINDAR-project: In order to support the interpretation of *brains* as DOMINDARs, macroscopic superposition states must be naturally given, in great numbers, within the natural environment of animals; it is not enough if they are here and there artificially produced in the lab. Judging from our human position, there *do seem* to be uncountably many naturally given states of macroscopic indetermination within the natural range of animals (especially humans), but there do not seem to be within the natural range of animals many naturally given macroscopic superposition states. This suggests that the connection between macroscopic states of indetermination and superposition states is not as close as initially hoped for. In the life-world, in the world in which animals have to make their decisions, the number of superposition states that are available for *incarnating* macroscopic states of indetermination appears to be just too small. Thus, it is far from obvious that quantum physical states of superposition contribute significantly to a solution of the detection-problem.

Perhaps the detecting of macroscopic indetermination by DOMINDARs is not literally a detecting; perhaps it is, properly speaking, an inferring, or even a postulating. If what is inferred or postulated turns out to be really there, at least in a great number of cases, then the inferring, respectively postulating, serves the purpose of obtaining reliable information not significantly worse than it would be served by *detection* in the literal sense. Taking this idea seriously, I propose that any state of *micro*-indetermination for a DOMINDAR—any occasion on which it is not predetermined which of its input-states occurs—is, in fact, *incarnated* by a state of quantum-physical superposition, a certain non-classical state of actuality; *whereas* any state of macro-indetermination for a DOMINDAR is, indeed, *merely* a state of pure possibility (*not* incarnated by a superposition state, a certain non-classical state of actuality). There is, after all, no problem in this, because under the presently considered hypothesis the states of macro-indetermination for a DOMINDAR do not need to be literally *detected* by it. It is true of the *usual* DOMINDAR that a state of micro-indetermination for it does not betoken with logical certainty that there is also, corresponding to that state, a state of *macro*-indetermination for the DOMINDAR[11] (that is, an occasion—in a certain temporal distance—on which it is not predetermined which of its output-states occurs); that the DOMINDAR's states of micro-indetermination are *incarnated* by states of quantum-physical superposition does not change this (usual) fact. But given the truth of the DOMINDAR-conditionals—each with an input-state of the DOMINDAR on the *protasis*-side and the corresponding out-

---

11 But we have seen what an exception to the *usual* DOMINDAR would look like: the SCREACTOR (if it were a DOMINDAR).

put-state on the *apodosis*-side, and taken together completely describing the determination relation between the DOMINDAR's input- and output-states—a state of micro-indetermination for the DOMINDAR is *an indication* of a state of macro-indetermination for it. Micro-indetermination for the DOMINDAR justifies *assuming* the corresponding macro-indetermination.

On the level of the logical bare bones: Though the following form of inference is not *logically* valid,

If A, then B. Not necessarily A → Not necessarily B,[12]

particular instances of this inference-form are, nevertheless, rationally useful, namely, as bases of *Peircean abductions* of a peculiar kind: *non-necessity abductions*. Consider Peter. If he wins in the lottery, he will buy a BMW. But of course it is not a necessity that he will win in the lottery. We infer (not in a logically valid way, but still quite justifiedly): It is (therefore) not a necessity that Peter will buy a BMW. Or consider the Geiger-counter. If this particular atom decays at t, then the Geiger-counter will click at $t+\delta$. But it is not a necessity that the atom decays at t. We infer: It is (therefore) not a necessity that the Geiger-counter will click at $t+\delta$. The inferential quality of a non-necessity abduction depends, of course, on the availability (more precisely speaking: the *extent* of the availability) of routes of necessitation that are viable *alternatives* to the route of necessitation presented by the first of the abductive inference's two premises, the premise with the form "If A, then B"—routes that might lead to the necessity of B *even in the absence* of the necessity of A. In the two examples of non-necessity abductions just adduced, the inferential quality is rather high (under normal circumstances, which we—quite automatically—assume to obtain), and in the second abduction still higher than in the first.

It seems to me that what I called "the detection of macro-indetermination" by a DOMINDAR is in fact the performance, by the DOMINDAR, of a non-necessity abduction, on the basis of its input-output conditionals and on the basis of a state of micro-indetermination for it. The abduction is fallible, but that does not mean that it is not reasonable. Having called it "reasonable", I immediately add that a DOMINDAR need not have any idea of *how* it arrives at postulating macro-inde-

---

12 This inference form and the inference form "If A, then B. Necessarily B → Necessarily A" are not logically valid (but are here understood in such a way as to be logically equivalent). In contrast, "If A, then B. Necessarily A → Necessarily B" and "If A, then B. Not necessarily B → Not necessarily A" are (understood in such a way as to be logically equivalent and) logically valid. In this paper, I have, in effect, made use of the latter two inference forms: cf. footnotes 7 and 8.

termination, let alone of how reasonable the procedure is. The DOMINDAR which is the human brain certainly has no idea of how it comes to conclude that there is macro-indetermination for it (in fact, it is entirely unaware of the inference); *a fortiori*, it has no idea of how reasonable, under the circumstances, are the non-necessity abductions it performs. It just implements the procedure. And it presents *us*—the human subjects of consciousness—*in the mode of consciousness* with the *conclusions* (*merely* with the conclusions, and those *not as* conclusions) of its automatically performed, objectively reasonable inferences: We have, normally, the consciousness of macro-indetermination *for us*—the feeling that we might do *this*, or alternatively *that*, the feeling of many possibilities open to us *now*; this consciousness steadily accompanies our waking hours of normal life. If this consciousness of freedom were always or usually untrue, *its insistent occurrence* would be quite unexplainable from the biological point of view.

Having spoken at some length about the detection-problem for DOMINDARs, I finally come to the determination-problem. Whereas the detection-problem concerns the states of macro-indetermination for a DOMINDAR, the determination-problem concerns the states of micro-indetermination for it. There is, obviously, no independent determination-problem concerning the states of *macro*-indetermination for a DOMINDAR, since a DOMINDAR restricts macro-indetermination *via*—only *via*—restricting micro-indetermination. Likewise, there is no independent detection-problem concerning the states of *micro*-indetermination for a DOMINDAR: a DOMINDAR does not detect these states, it is simply *in* these states (or rather, some microscopic part of it), and on their basis it "detects"—that is: implements an abductive inference of the existence of—the corresponding states of macro-indetermination. I propose (see above) that the states of micro-indetermination for a DOMINDAR—in themselves states of pure possibility—are *incarnated* in states of actuality: *in states of quantum-physical superposition*. If so, a DOMINDAR's abolishing of micro-indetermination is not a sort of *creatio ex nihilo*, not even a *creatio* out of the moderate *nihilum* of pure possibility. Rather, that abolishing is *like* the making-clear, the making-precise of a vague term, of a term that can be made clear or precise in more than one way; in other words, it is more like a decision than a creation. Clearly, quantum physics contributes crucially to avoiding the appearance that the work of a DOMINDAR must be something like the creative work of God.

But the center of the determination-problem is that superpositions seem to mutate into definite states by pure chance. If there is, on a certain occasion, a state of micro-indetermination for a DOMINDAR and it is not predetermined which of the several input-states of the DOMINDAR occurs at time t, then each of these input-states has, on the given occasion and as an ingredient in a quantum-physical superposition state, a certain objective probability of occurrence at *t*. Suppose it is predetermined that one of the DOMINDAR's input-states occurs at *t*—although it is not predetermined *which* of the DOMINDAR's input-states occurs at *t*. Then the mentioned probabilities add up to 1, for occurrence at *t*—though none of them *is* 1 for occurrence at *t*. *Later*, the probabilities of the input-states still add up to 1, for occurrence at *t*, but now *one* of the probabilities *is* 1 for occurrence at *t* (and all the other ones are, therefore, 0). The superposition state has mutated into definiteness, in other words, *one* of the microscopic input-states of the DOMINDAR has now been *determined* to occur at *t* (and accordingly, given that the DOMINDAR is well-functioning, one of its macroscopic output-states has been determined to occur at *t+δ*). The problem is that saying "this input-state has now been determined to occur" seems a purely metaphorical way of speaking; for there appears to have been *no one and nothing* that did the determining. The DOMINDAR, certainly, appears to have nothing to do, causally, with the mutations into definiteness of the superposition states that incarnate its states of micro-indetermination. These mutations appear to occur by pure chance, the DOMINDAR having no control whatsoever over them.

If this turned out to be the true general state of the matter, then there would be no DOMINDARs *in the strict sense*, all DOMINDARs (i.e., the things that one nevertheless calls "DOMINDARs", perhaps out of habit) would be DOMINDARs *only in an attenuated sense*. For they would have no *determiner* in the true sense of the word;[13] their so-called "determiner" with respect to microscopic indetermination (and hence also macroscopic indetermination) would be *pure chance*. DOMINDARs have already been found to have no *detector* in the true sense of the word; their so-called "detector" of macroscopic indetermination is not really a detector but, so to speak, an *inferential postulator*. Already for this reason alone, all DOMINDARs are DOMINDARs only in an attenuated sense (keeping in mind what "DOMINDAR" is an acronym for). But the *second* extension (or thinning) of the meaning of the term "DOMINDAR", now threatening, would be considerably more problematic than the first. Recall that I asserted that it is impossible to make a DOMINDAR out of the SCREACTOR because there is no known way to *determine* that a radium atom decays, or does not decay, at time *t*. Clearly, in asserting this, I implicitly excluded pure (objective) chance from being a way of *determining* that a radium atom decays, or does not decay, at a given time. I had good reason for

---

[13] I am here using terminology that I introduced only for two-ways DOMINDARs for formulating conclusions that concern all DOMINDARs. Note that for more-than-two-ways DOMINDARs, "restrictor" is more appropriate than "determiner"; for in their case, *restriction*—i.e., excluding a possibility or some possibilities from actualization—is not automatically *determination*, that is, picking precisely one possibility for actualization.

this exclusion: pure chance ought not to be considered a selective agent, and it ought not to be considered a causal agent. But now it seems that pure chance has to be *included* among the possible *determiners* and that the notion of a *determiner* has to be stretched in this manner, so that at least *in an attenuated sense of the term* "DOMINDAR" there can be DOMINDARs. And pure chance may well be the only determiner—"determiner" (in scare quotes)—that *can be* "put to work" in a prospective DOMINDAR. *Chance*-DOMINDARs may well be, by natural necessity, the only DOMINDARs there can be.

That *chance*-DOMINDARs are necessarily the only DOMINDARs—this is what many would accept as the ultimate verdict of physics. But, so far, I refuse to concur; for from the biological point of view *such* an ultimate verdict of physics would be highly unsatisfactory. For it seems undeniable: If the human brain—a product of biological evolution—is a DOMINDAR, then it is not a chance-DOMINDAR but a *rational* DOMINDAR, a DOMINDAR that restricts macro-indetermination *not blindly* but so as to fit means to ends. If DOMINDARs are favored by biological evolution, as they certainly seem to be, evolution leading to ever higher developments and sophistications of the basic DOMINDAR-model (the present apex is, as far as we know, the human brain, which even produces reflexive consciousness), then it can hardly be true that all DOMINDARs are—let alone: must be—*chance*-DOMINDARs. For the biological usefulness of a chance-DOMINDAR—that is, its contribution to the survival capacity of the organism that is equipped with it—is severely limited. A chance-DOMINDAR can serve as a tie-breaker in situations where it is not important what (among the given alternatives) is done—where it is only important that something (among the given alternatives) *be* done. A chance-DOMINDAR will certainly save Buridan's Ass from dying of starvation. But it will only accidentally—only by chance—help the poor animal to escape from the lion's maw.

Chance-DOMINDARs are DOMINDARs in a broadened sense of the term, but they simply do not have enough of DOMINDAR-hood to fit all the facts of evolutionary biology. However, at this point I must confess my ignorance. I do not know how a DOMINDAR that is *more than* a chance-DOMINDAR is physically realized or realizable. Here we have the truly hard problem for the DOMINDAR-project. It is not enough that evolutionary biology has, quite plausibly, a place for DOMINDARs that are more than chance-DOMINDARs; it would seem that the physical implementation of such devices must also be described in a true and convincing way. The wherewithal for a solution of the problem, assuming that it has a solution, may be expected to be found in certain, not yet clearly ascertainable quantum-physical features of the brain. (Where else might it be found?) Along with others, I *surmise* that those features are closely connected to, perhaps identical with, certain hypothesized quantum-physical features of the brain which, supposedly, enable it to bring forth consciousness. But this—not very precise—*surmise* is all I have.

In order to dispel an impression that my ignorance is merely personal, consider the quantum-theoretical model of human action proposed by Henry Stapp. It is quite sufficient to consider Stapp's approach in its most general terms.[14] According to Stapp (utilizing quantum-theoretical ideas of John von Neumann's), a human action—say, raising one's arm—has two aspects, the conscious intention and the physical action linked to it, and it is determined by four processes (some of which, in their turn, deserve the name "action", some of which don't): process 0, process 1, process 2, process 3. Following von Neumann's terminology, process 2 is the undisturbed evolution of the physical system, in accordance with the Schrödinger equation, and process 1 "the basic probing action that partitions a potential continuum of physically described possibilities into a (countable) set of empirically recognizable alternative possibilities" (Stapp 2011, 24). Process 1 is an *intervention* in process 2, an intervention inexplicable by the formalism of quantum theory, but nonetheless necessary if quantum theory is to have empirical import.[15] Process 0 (Stapp's terminology) is the partly conscious selection process, inexplicable by the formalism of quantum theory, which determines the "process 1 action", the "basic probing action" just described. Process 3 (Stapp's terminology), finally, "*selects the outcome*, 'Yes' or 'No', of the probing action. Dirac calls this intervention a 'choice on the part of nature', and it is subject, according to quantum theory, to statistical rules specified by the theory" (24; the emphasis is Stapp's). This account is likely to leave the reader with the impression that the ultimate and decisive agent of a human action is, according to Stapp, *not* the human being to whom the action is ascribed, and *not* the soul or subject of consciousness of that human being—and Stapp could hardly disagree, given his claim "I introduce no ghosts" and his endorsement of William James's dictum "The thought itself is the thinker" (133),[16] not to speak of Stapp's favoring of Whitehead's anti-substance process ontology (see Stapp 2011, 85–98). However, Stapp's account is also likely to leave the reader with another impression, an impression that is certainly contrary to Stapp's best intentions: the impression

---

14 For what follows, see Stapp 2011, 23–24.
15 "[T]he orthodox formulation of quantum theory [...] asserts that, in order to connect adequately the mathematically described state of a physical system to human experience, there must be an abrupt *intervention* in the otherwise smoothly evolving mathematically described state of that system" (Stapp 2011, 22; the emphasis is Stapp's).
16 Stapp's "No mental substance!" position is also apparent in Stapp 2009, 21–22. For Stapp, the thought is the *thinker*—and the actualization, Stapp seems to suggest, is the *actualizer*: "Suppose the actualized state of the brain is really *actualized*. What can this mean? One possibility is that some characteristic feature of this state becomes an actual 'experience'" (165; the emphasis is Stapp's).

that the ultimate and decisive agent of a human action is even *not* the mental process of the human being to whom the action is ascribed, but *physical nature*, which (in process 3) certainly acts to some extent *by pure chance* (non-statistical single-case physical necessity being out of the question in the orthodox quantum theory adhered to by Stapp). Stapp's analysis strongly suggests that all a human being's mental process *really does* (in constituting process 0 and in determining process 1) is "to set the stage" for *the action* (i.e., the human act) by putting in place, preliminary to the action, a range of distinct alternative possibilities (paradigmatically, two of them). The all-important rest is up to (physical) *nature*, in other words, up to (physical, objective) *chance*—which means that, in the end, the *intention of the agent* is quite irrelevant.

In the terminology of the present paper, Stapp's account of human action can be soberly summed up as follows: The intending agent determines a particular state of (physical) indetermination for itself, a particular set of alternative possibilities; but the action itself, in which precisely one of these alternatives is actualized, is *not* up to the intending agent; it just happens (with no explanation possible that goes beyond a merely statistical explanation, assigning probabilities). Now, a man, or his soul, or his mental process would not grudge *nature* the privilege of actualization if he, or his soul, or his mental process retained the right to *select* before actualization (or perhaps simultaneous with it) *the possibility that is to be actualized*. But according to Stapp's account, the intending agent—which is for Stapp the mental process of the human being—does certainly *not* retain that right. For according to Stapp *nature* not only actualizes but also *selects* the outcome, i.e., *the action*, the human act. The selection of the outcome is, according to Stapp, using Dirac's words (see above), a "choice on the part of nature", in other words: a "choice" on the part of *chance*—and the scare quotes around "choice" are quite justified (for *chance* is not only *causeless* but also *blind*).

Thus, Stapp's quantum-theoretical approach to human action offers no perspective to solve the determination-problem for DOMINDARs in such a way as to give DOMINDARs that are *more than* chance-DOMINDARs a substantial chance. The so-called *quantum Zeno effect*, invoked by Stapp in an attempt to get from mere "probing actions" and "choices on the part of nature" to *intentional actions*,[17] is of no considerable help. The quantum Zeno effect consists in

---

17 Stapp *is* sensing a problem for his account: "But the only dynamical freedom offered by the quantum formalism in this situation is the freedom to perform at a selected time some process 1 action. Whether or not the 'Yes' component is actualized is determined by 'nature' on the basis of a statistical law. So the effectiveness of the 'free choice' of this process 1 in achieving the desired end would generally be quite limited. The net effect of this 'free choice' would tend to be nullified

the following phenomenon (see Stapp 2011, 35–36): If a process 1 action X, with a particular process 3 outcome, is the first element in a very rapid sequence of process 1 actions very similar to X, then all the process 1 actions in that sequence will, with high probability, have the same kind of process 3 outcome as X had. Stapp believes that the rapidity of a sequence of "essentially identical" process 1 actions—and therefore the likelihood of the Zeno effect—can be increased by a mental effort of attention (36–37). Therefore, according to Stapp, the intending agent of a human action (which agent is for Stapp—to repeat—the human being's mental process) has *some* power to hold nature, or chance, to *its word*, i.e., to its original choice—at a time when the word has already been spoken, the choice already been made! One may well ask: Is *this*—just *this*—supposed to be what the *intendedness* of a human action consists in? The mere mental insistence on an outcome *chosen by nature* (but is it even that?, or is it merely the mental insistence on a certain kind of "essentially identical" probing actions?), even if causally effective, seems altogether insufficient for *intending* that outcome. If *nature's choice* is not to the intending agent's liking, what then?—Then presumably the intending agent may by a mental effort of non-attention contribute to rendering nature's choice ineffective in the end? (But Stapp, as far as I can see, is silent on this issue.)

Given the central (but in its centrality not quite acknowledged) role that objective *chance, randomness*, plays in Stapp's account of human action[18]—even *with* the quantum Zeno effect in place—it is clear that one cannot profit from that account if one wishes to recognize DOMINDARs that are, truly, *more than* chance-DOMINDARs. The existence of DOMINDARs—indeed, the existence of *non-idle, active* DOMINDARs—is called for by our self-experience and by evolutionary biology, and it is consistent with quantum physics (though not with classical physics). The existence of active DOMINDARs is even consistent with the causal closure of the physical—*if* chance-DOMINDARs are all the DOMIN-

---

by the randomness in nature's choice between 'Yes' and its negation 'No'" (Stapp 2011, 35).

18 At one point Stapp touches on the problem: "The advance to quantum theory appears at first to offer no basis for any significant improvement: choice is now distributed over time, [...] but is asserted to be controlled exclusively by 'pure chance'" (Stapp 2009, 169). Choice *would* be controlled by pure chance if nature were the ultimate chooser. And that nature *is* the ultimate chooser seems to be what Stapp's view ultimately comes down to. It does not help to *call* nature's choices (for no good reason) "*intrinsically meaningful*: each quantum choice injects meaning, in the form of enduring structure, into the physical universe" (169; the emphasis is Stapp's). Enduring structure is certainly not *per se* a form of intrinsic meaning: there can be plenty of enduring structure in a physical universe that is absolutely meaningless. What is true, however, is that enduring structure is a *necessary condition* of whatever meaning there is in a physical universe.

DARs there are.[19] But in fact our self-experience and evolutionary biology demand active DOMINDARs that are *more than* chance-DOMINDARs: they demand active DOMINDARs that are *rational* (which as such, it seems, have to be *instrumental* DOMINDARs—instruments for *conscious souls*).[20] The existence of active rational DOMINDARs is still consistent with quantum physics (though hardly with materialism); it is, however, an open question whether quantum physics can significantly help us with the determination-problem for *such* DOMINDARs. They are, as *rational* DOMINDARs, DOMINDARs not only in the broad—the attenuated—sense but also DOMINDARs *in sensu stricto*, since they require *determiners* in the true sense. Chance—which is blind and, properly speaking, not a causal agent at all—is certainly not "determiner" enough for *such* DOMINDARs. But how would their determiners *work*? Will we ever know?

---

19 The mere admission of physical chance-events—physical events without sufficient cause—does not hurt the causal closure of the physical; it only hurts physical determinism.
20 Stapp, not a friend of substantial souls, at least quotes and endorses William James, who in his *Principles of Psychology* forcefully argued on biological grounds for the causal efficaciousness of consciousness (though not for the causal efficaciousness of its *enduring subject*). (See Stapp 2011, 3–4; Stapp 2009, 10–11.)

References

Meixner, U. 2004, *The Two Sides of Being*, Paderborn: Mentis.
Meixner, U. 2006, "Consciousness and Freedom", in A. Corradini, S. Galvan, and E.J. Lowe (eds.), *Analytic Philosophy Without Naturalism*, London: Routledge, 183–196.
Meixner, U. 2008, "New Perspectives for a Dualistic Conception of Mental Causation", *Journal of Consciousness Studies* 15 (2008): 17–38.
Meixner, U. 2010, "The Emergence of Rational Souls", in A. Corradini and T. O'Connor (eds.), *Emergence in Science and Philosophy*, New York/London: Routledge/Taylor & Francis, 163–179.
Stapp, H.P. 2009, *Mind, Matter and Quantum Mechanics*, Third Edition, Berlin/Heidelberg: Springer.
Stapp, H.P. 2011, *Mindful Universe. Quantum Mechanics and the Participating Observer*, Second Edition, Berlin/Heidelberg: Springer.

Godehard Brüntrup
Quantum Mechanics and Intentionality

1 Two mysteries—loosely connected?

The Authors who have suggested a deep connection between the solution of the mind-body problem and the understanding of quantum mechanics are too numerous to list. But what exactly would be the alleged structural fit between these two deeply puzzling theoretical issues? Can a genuine philosophical issue—which by its very nature is conceptual in nature—ultimately be resolved by advances in the empirical sciences? The connection is all too often asserted to lie in the subjects' confounding natures, but the mere fact that two problems are of equal impenetrability to the human mind does not imply that there is a profound interdependence between the two. There is however an obvious way in which quantum mechanics might have a bearing on the philosophical mind-body problem: if quantum mechanics implies that for some physical events there is no physical cause, then at least the very strong variants of the so-called "principle of causal closure" of the physical realm lose much of their corroborative underpinning in physics as practiced by real physicists as such, and are thus constrained to the speculative physicalism advanced by metaphysicians. However, unless one assumes a strong version of the principle of sufficient reason, which would exclude chance events in principle, even events supposedly not caused by some physical cause are not immediately thereby caused by some *non*-physical cause. Moreover the fact is that not all of the interpretations of quantum mechanics imply indeterministic 'choices' of events: neither the many-worlds interpretation nor a Bohmian hidden variable interpretation assume indeterminism. Even in the most obvious theoretical nexus between quantum mechanics and philosophy of mind, the theoretical gain is much smaller than initially thought. Our understanding of the ontology at a microphysical level is indeed so limited that just about any position in the philosophy of mind can be construed in such a way that it is in agreement with quantum mechanics, at least in one of its possible understandings. There is, however, one philosophical concept that has quite significant bearing on quite a number (but not all) interpretations of quantum mechanics. It is the concept of intentionality. In this paper it will be argued that the concept of intentionality is the most promising theoretical bridge between the interpretation of quantum mechanics and the philosophy of mind. How quantum mechanics is connected to consciousness is in the end dependent on how we think that consciousness is related to intentionality.

2 The seemingly incoherent world of quantum mechanics

The common understanding of the problem surrounding quantum mechanics is one of interference: the investigated phenomena are so small that it becomes unavoidable to not change their properties during measurement. The observer is never a passively reporting entity, but always a reality-changing interfering entity. The real problem of quantum mechanics is also not yet fully captured by stating that particles behave like waves when unobserved. While this is true in a very general way, it does not yet encompass the full gravity of the problem that the very nature of the object before measurement is such that we cannot comprehend it. The objects of quantum mechanics do not have the same identity criteria as classical concrete entities. For example, if we have two boxes and two classical objects A and B, the state whereby A is in the left box and B is in the right box is distinct from the state whereby the two objects are switched such that B is in the left box and A is in the right box. In quantum mechanics these supposedly different states are indistinguishable. When not measured, quantum objects are in a strange state that is yet to be fully understood. It is called 'superposition', and dynamics of these states can be calculated with the robust and well-established 'Schrödinger equation', but understanding what it means to be in this state has proved elusive to such an extent that it might be due to a limitation of human understanding, i.e. a Kantian boundary of reason, as the behavior of particles in this state seems to defy the laws of logic. In a variation of the famous two-slit experiment, it can be shown that particles capable of traveling through exactly two possible pathways in an experimental setup follow neither the first, second, both paths, or no path, yet these are the only possibilities allowed for in classical thinking. That the particles do not take both paths can be shown by interrupting the experiment, which leads to the discovery of the particle in one of the two pathways, but that the particles do not travel through one path only, can be shown by blocking one of the two pathways. This in effect changes the statistical distribution of properties of the particles leaving the apparatus, compared to the original case where both pathways were open. That the particles do not travel via some other route can be demonstrated by blocking both pathways, which results in no particles leaving the apparatus whatsoever.

A detailed description of this experiment was published by David Albert (Albert 1994), but the technical details are at this point not necessary to comprehend the predicament. Some have suggested that we need to revise the laws of logic to make room for the behavior of quantum particles, but even that would not help us in understanding what is going on in this case, as there are other features of quantum behavior that defy human understanding such as non-locality. Particles entangled in a superposition behave somehow as if they were only one thing. Measuring one entangled particle has immediate effects on far removed particles with which it is entangled without any time delay, despite massive distances in between them. While this behavior is certainly not excluded by standard logic, it seems to be inconsistent with the intuition that the world is made up from distinct particulars. The problem with quantum mechanics is thus not simply the interference of the observer with the observed objects but that the very nature of the quantum realm seems bizarre and incomprehensible to us.

Three basic theoretical frameworks have been given to account for these difficulties in grasping the ontology of the quantum world. In order to investigate possible relations to the philosophy of mind, we first have to briefly delineate the logical map of these competing interpretations of quantum mechanics. One way of laying out this logical space is by constructing a trilemma. Here, of three possible claims only two can be mutually consistent, leading to three possible combinatorial solutions. The three claims are:

(I) The dynamics of the system is completely governed by the fundamental Schrödinger equation. There are no additional indeterministic processes.
(II) Our knowledge of the system is essentially complete. There are no hidden variables.
(III) A measurement yields a unique result at the expense of other nomologically possible results. There is no branching of the universe.

From a scientific point of view it is obvious why one would want all three claims to be true: a universe governed by a deterministic equation without random interference is certainly preferable for the sake of making predictions, the idea that we have not missed something of crucial importance is equally attractive, and that the outcome of a measurement is informative only if it happens as exclusive to other possible outcomes. Quantum mechanics forces us, however, to drop one of the three claims.

If (I) the Schrödinger equation is all there is and we have (II) not overlooked something important, then we do not get unique measurement results (not III). The many-worlds or many-minds interpretation bites this bullet and claims that all possible outcomes are realized, each relative to a world or a mind (not III). The standard interpretation of quantum mechanics, however, denies (I). In addition to the process described by the Schrödinger equation there is an indeterministic process by way of which one of the possibilities within the realm given by the Schrödinger equation is selected. Thus the wave package collapses into one determinate result, what is often referred to as the 'collapse interpretation'. A negation of (II) introduces additional entities and mechanisms, the best-known case of which is arguably Bohm's interpretation of quantum mechanics (Goldstein 2013). There is no collapse in this theory and the wave function does not represent worlds or minds, but is considered to be something like a pilot wave that directs the movements of the particles via the "active information" contained in it.

Each of the three major interpretations offers a strikingly different ontology. Unless future advances in experimental design and theoretical understanding provide a clearer picture of the ontology of the quantum world, the situation will remain one in which philosophers of mind cherry-pick the interpretation that best fits their favorite ontology, which is in fact the situation today. Unless it changes, the contribution of quantum mechanics in understanding the mind-body problem will remain limited. Empirical theories function as test cases for metaphysical theories. If the philosophical theory is in obvious disagreement with the empirical facts, it has to give way. But if the empirical facts are but a mathematical formalism which yields reliable predictions but whose ontological implications are completely up for grabs, then it cannot serve well as a something that limits and guides philosophical speculation. This is exactly the predicament in which we find ourselves when we try to establish connections between quantum mechanics and the mind-body problem in philosophy of mind.

3 The Mind-Body Problem

At first glance the mind-body problem seems to be a bundle of different but related problems in metaphysics. The question of the causal role of the mental in the physical world ('mental causation') and the closely connected question problem of free will are probably the most intuitive and widely known versions of the mind-body problem. Philosophically it seems that the fundamental issues underlying all other aspects of the mind-body problem are intentionality and phenomenal consciousness. It must be said that these two concepts might be so closely connected that they are ultimately only two sides of one coin, but for the sake of conceptual clarity and to advance the thesis of this paper we must keep them separated. The argumentative thrust of this paper is, after all, that in some interpretations the concept of intentionality will take up a paramount role relevant to the understanding of quantum mechanics, and that consciousness will come into play via a relation to intentionality.

Intentionality needs to be understood as the 'aboutness' of mental content. When we think, we think *about* something, when we make a statement, it contains propositional content, *about* which this statement is. The contemporary discussion of intentionality probably started with Franz Brentano (1838-1917) who stated that every mental phenomenon includes something as object within itself, i.e. in presentation something is presented while in judgment something is affirmed or denied. The deep philosophical question is how aboutness or intentionality could be possible in a physical world. It seems to be different from physical relations like causation because intentionality can be directed towards possible or non-existent states of affairs. It seems that the connection between the mind and what it represents cannot be captured in physical categories.

The other fundamental version of the mind-body problem is the puzzle how phenomenal experience can arise in a physical world. The so-called 'hard problem of consciousness' (David Chalmers) asks this central question: even when we have explained the performance of all the cognitive and behavioral functions, why is the performance of these functions accompanied by conscious experience? In other words: it seems conceivable that there are other possible worlds which contain creatures that behave just like us, process information just like us, interact with the environment just like us, but don't possess even the faintest internal 'light' of consciousness. They would be perfect humanoid robots or, as philosophers like to call them, 'metaphysical zombies'.

The question of how intentionality and phenomenal consciousness are related is complex and there is significant disagreement among philosophers on this issue. Are all intentional states conscious or at least potentially conscious? Are there intentional states that will never become conscious? For example, medical patients who experience blindsight seem to represent mental content without ever being consciously aware of their own intentional state. There is subliminal processing of symbols and informational content, which never reaches the threshold of conscious experience. Vast arrays of mental operations happen in a subconscious or unconscious realm, but they nevertheless are full of representational content. They thus can be classified as intentional states exhibiting "aboutness" in the relevant sense mentioned above. Relatedly, are all conscious states intentional states? There might be raw feelings that do not represent anything outside of themselves, that is they have no connection with anything outside of the mind itself. For example, in German one distinguishes the raw feeling of "Angst" (anxiety) from the directed feeling of "Furcht" (fear). *Angst* has no intentional object while fear has such an object.

Even these few observations make it perfectly clear that the relationship between intentionality and phenomenal consciousness is open to interpretation. Any serious theoretical account of the mind-body problem has to take a stance on this issue of the interdependence of intentionality and phenomenal consciousness. The relevance of quantum mechanics to the mind-body problem is dependent on how one construes this relationship.

4 The relationship between intentionality and consciousness

Four accounts of this relationship can be distinguished (cf. Siewert 1998):
(1) *Consciousness derived—Intentionalizing consciousness*: Consciousness is explanatorily derived from intentionality. This strategy takes intentionality as basic and construes conscious states from intentional states.
(2) *Consciousness separable—Consciousness as non-intentional raw feeling*: Consciousness is separable from intentionality and cannot be derived from intentionality. Phenomenal states are conceived of with respect to their qualitative content (qualia) and raw feelings are void of intentional aboutness.
(3) *Consciousness inseparable—Phenomenal consciousness is sufficient but not necessary for intentionality*: Consciousness is not derived and inseparable from intentionality. Wherever there is consciousness there is intentionality but not vice versa.
(4) *Consciousness necessary—Strong modal tie between consciousness and intentionality*: Consciousness is not derived from, inseparable from, and essential to intentionality. All intentional states are conscious states.

It is only with these distinctions in mind that we can get a better view on how exactly quantum mechanics might be relevant to philosophy of mind. It seems unlikely and conceptually difficult that quantum mechanics is directly connected with phenomenal consciousness. This is so for broadly Cartesian reasons: it is hard to build a conceptual bridge between something mechanical and phenomenal qualia. Phenomenal content is defined by its intrinsic quality and not by being related to other entities. Intentionality is, however, a relational notion. The chances of being able to connect it somehow to the mechanical structure of the world, that is the causal network, seem higher from the outset. For this reason we will make a crucial decision at this point. We will first try to connect quantum mechanics to intentionality and only then ask the additional question how the gap to phenomenal consciousness might be bridged. The key insight that will be defended here is *that quantum mechanics might give us some help in understanding how intentionality is rooted in physical reality, because quantum mechanics suggests there is some form of representation and information processing built into the very fabric of the universe*. If that is the case, then account (1) and (4) seem to be the most promising for understanding consciousness as well, while accounts (2) and (3) are less feasible. The reasons for this are obvious: in (2) consciousness is separable from intentionality and as such not representational but a 'raw feeling', whereby quantum mechanics cannot offer much insight into understanding consciousness. In (3) consciousness is sufficient but not necessary for intentionality and all conscious states are intentional, but only some intentional states are conscious, and consciousness can not be derived from intentionality. Here quantum mechanics does not help explain why some intentional states are conscious, and others are not. Account (1) is more promising: if quantum mechanics makes reference to intentionality at the fundamental level of the universe, then the emergence of higher levels of intentionality, and thus consciousness, is less problematic. It could be the case that conscious higher-order intentional states supervene on unconscious lower-level intentional states. A higher order theory of consciousness (HOT) could accommodate this idea. In this case physicalism might be true, if the supervenience relation is strong and intentionality can be construed as bona fide physical (causal?) relation. There are however other non-physicalist versions of (1).

In account (4), i.e. where there is a strong modal tie between consciousness and intentionality, there also seems to be a promising solution. Assuming that quantum mechanics shows that some form of intentionality (representation of informational content) is built into the very fabric of the physical universe, then, if by some form of mutual modal entailment all intentional states are conscious and quantum mechanics cannot be formulated without reference to intentionality, assertions of some kind of panpsychism or pan(proto)psychism might be true.

In short we get the following picture of the four accounts if we construe our interpretation of quantum mechanics in such a way that it locates some form of (proto)intentionality in nature:

(1) *Consciousness derived*: Grounding intentionality in the physical world is a reliable basis to understand the emergence of consciousness as a higher-level intentionality (HOT account of consciousness). Quantum mechanics might be helpful in explaining consciousness.

(2) *Consciousness separable:* Grounding intentionality in the physical world will not help in understanding consciousness if it is a distinct phenomenon, i.e. intrinsic, raw feeling without intentional content. Quantum mechanics is not in an explanatorily relevant sense connected to consciousness.

(3) *Consciousness inseparable*: Even though all phenomenal states are intentional states (inseparable), intentionality does not require consciousness. Neither is understanding intentionality sufficient for understanding consciousness, nor is grounding intentionality in the physical world helpful in understanding consciousness. Quantum mechanics is only accidentally and not in an explanatorily relevant sense connected to consciousness.

(4) *Consciousness necessary*: If consciousness is essential to all intentional states, it follows that: if intentionality is a fundamental feature of nature, then so is consciousness. Grounding intentionality in the physical world

results in grounding consciousness in the physical world. Quantum mechanics is essentially connected to consciousness.

It is now time to put our working hypothesis to the test. To this end we will look at the three major interpretations of quantum mechanics as they appear given the denial of one of the three horns of the trilemma above.

5 Collapse theories and intentionality

Denying (I): It is not the case that the dynamics of the quantum system are completely governed by the fundamental Schrödinger equation. There are no additional indeterministic processes.

Various versions of collapse theories fall under this heading. They introduce an additional indeterministic process or event: the collapse of the wave function. The assumption of a collapse or reduction seems like a crude *ad hoc* assumption to "force ontology to kneel to prejudice" (Stapp 1989, 157). Indeed, the by now classic GRW version of the collapse theory (Ghirardi, Rimini, and Weber 1985) is ultimately such a brute assumption: each elementary particle is subjected to random and spontaneous localization processes. And even though this can be perfectly described mathematically, there is the inexplicability of mere chance that renders this entire process a somewhat theoretically unsatisfying *ad hoc* construct. Since these spontaneous reductions are not conceptually connected to any mental activity, there is little theoretical gain to be made here for the philosophy of mind. Henry Stapp's version of a collapse theory is quite different (Stapp 2007). For him quantum reality, the superposition, collapses into classical reality if *probed for a specific information* (cf. Heisenberg's Uncertainty Principle). If a question is asked, for example if measured for a specific spin, the system provides specific information in return. From a philosophical perspective the crucial point here is that the measurement process as *asking for a specific information* may be seen as something implying intentionality. In that the quantum system is being represented as specific informational content, a meaningful question posed to the system can thus also be represented. Now, unless these expressions are mere metaphors and the real process is simply nothing other than a random collapse, there is some intentionality built into nature wherever these collapses occur. It is not ultimately relevant at what level of nature these probing actions happen; maybe they happen all the time even at the lowest level, or possibly they require some higher form of mentality. In any case they require something that goes beyond a classical mechanism: that nature answers a question posed to the system requires that a *decision* be made as to what the question will be. According to Stapp, in the collapse a big smear of possibilities is being reduced, and this reduction increases knowledge: by becoming more determinate and realizing possibilities, information is represented in the universe, which in turn can be used to realize new possibilities consistent with what already has happened and what is possible by the laws of nature in the future. The representation of information is thus the key idea in Stapp's theory; for Stapp quantum theory sees the physical world in terms of information. The billiard ball view of classical materialist atomism is replaced by the notion of a holistic and at least partly nonmaterial world consisting of an objective carrier of a growing collection of non-localized bits of information. Representation of information, however, is a form of aboutness, it implies some form of intentionality. The crucial philosophical question is whether such intentionality requires consciousness or experience. If we follow account (1) (consciousness derived from intentionality) then a vast quantity of intentional acts of representation will happen without consciousness, and only some very complex ones will be accompanied by a moment of full phenomenal consciousness. Following Whitehead's process metaphysics, Stapp sees the world as dynamically related experiential events. But for Whitehead experience does not necessarily imply consciousness, that is to say that all conscious events are experiences but not all experiences are fully conscious. So, Stapp's view is in principle open to an interpretation in which some form of non-conscious intentionality (Whitehead calls it 'prehension') permeates the universe. Another interpretation might be that the representation of information in a collapse occurs only with respect to fully conscious observers. This would be more in line with account (4), where a strong modal tie between intentionality and consciousness is claimed. Where there is intentionality, there is consciousness. It is not crucially important here to decide this question of interpretation here. What is important is that our hypothesis is corroborated: it is via the notion of intentionality that quantum mechanics becomes relevant for the philosophy of mind.

Michael Epperson's account is another recent philosophical interpretation of quantum mechanics inspired by Whitehead (Epperson 2004). Epperson makes extensive use of the concept of "decoherence" and interprets it as negative selection. The concept of decoherence is more often used by many-minds or many-worlds views. His theory is thus not a classical collapse account, but nevertheless not a many-minds or many-worlds interpretation. In the process of negative selection the coherent multiplicity of relations is reduced to a set of decoherent and mutually exclusive potential novel integrations. This process guarantees that the history of entities is mutually consistent, and the possible future paths of the universe are consistent with the actual past. At the end of this process, one particular path is chosen from those that are consistent with the actual past. Here Epperson does not take the route of the many-world or many-mind interpretations: only one outcome becomes actual. Thus we have a constant flow from actuality to potentiality to new actuality. But this process requires the *representation* of all possibilities, a calculation of those that are consistent with the past and the states of all other entangled entities. At each moment every physical event at the quantum level brings about its successor by calculating, evaluating and reducing possible future states given the past history of every other event with which it is connected by virtue of quantum entanglement. It is the probability valuation of the mutually exclusive states that governs the actualization of a unique outcome state (Epperson 2004, 102). It is not necessary to venture more deeply into Epperson's fascinating interpretation and how it relates to Whitehead's notion of 'concrescence'. The salient fact is obvious: at the quantum level not only actual events but also merely possible events and their logical relations are represented. Representation of mere potentialities is a paradigm case of intentionality. If representation is happening at the quantum level, then some kind of proto-mentality seems to be built into the very fabric of the universe. Whether consciousness is required for this kind of basic intentionality in the universe is again dependent upon the strategy one favors, that is whether one favors account (1) or account (4). Epperson does not mention consciousness at all; the representation and validation of yet unrealized possibilities seems to be happening without any form of consciousness involved.

Another form of the collapse theory was proposed by Penrose and Hameroff (Penrose, and Hameroff 2011). It is an objective collapse theory inspired by quantum gravity theory, whereby the quantum state remains in superposition until the difference of space-time curvature reaches a threshold. This happens all the way down at the minute Planck level of the universe. Larger objects are thus inevitably in a classical state. The thesis of Penrose and Hameroff with regard to the mind-body problem is this: each objective collapse is *identical* with a moment of proto-conscious experience, and larger orchestrated reductions are identical with moments of consciousness. Here conscious experience is simply identical with the collapse of the wave function. This claim articulates a psycho-physical identity theory and in a way leaves no room for further philosophical questions. Identities are brute and rock-bottom and don't allow for "deeper" explanations. Consciousness is not derived from anything else; it is not a higher-level phenomenon. It is built into the Planck level of the universe. This theory certainly entails some kind of panpsychism. According to Penrose, conscious systems are capable of intentional representation in such a way that is not possible for an algorithmic computational system to replicate their representations. Consciousness and at least certain forms of intentionality are closely connected. It seems to be a case of account (4), because it posits a strong modal tie between consciousness and full intentionality. But that is not entirely clear. In order to be more helpful in resolving the mind-body problem the Penrose-Hameroff theory ought to move beyond the claim of brute identities and develop stronger conceptual ties between quantum mechanics and intentionality.

6 Hidden variable theories and intentionality

Denying (II): It is not the case, that our knowledge of the system is essentially complete. There are hidden variables.

Bohm's ontological interpretation of quantum mechanics (Bohm, and Hiley 1993) is the best known example of a hidden variable theory, he in fact used the term in an early paper (Bohm 1952). In addition to the wave function of merely possible states it postulates a pilot wave that exists even when unobserved. At any moment there exists not only a wave function but also a well-defined configuration of the entire universe. The pilot wave guides the particles and informs them about the state of the entire system, ultimately the entire universe. So, for example, in a two-slit experiment, the particle, which passes through one slit, receives information from the pilot wave as to whether or not the other slit is open. Its trajectory is chosen in accordance with this information. Bohm called this kind of information 'active information' because the *content* of the information is what is truly causally relevant for the movement of the particle. Each particle has a rich inner structure that enables it to represent the information provided by the pilot wave and react to it (Bohm, and Hiley 1993, 37). "It is thus implied that in some sense a rudimentary mind-like quality is present even at the level of particle physics" (Bohm 1990, 283). An analogy would be a ship that is guided via a GPS-satellite. The information that is received by the ship is actively relevant for its movement. But it is not a physical force or field that is pushing the ship around. In fact, whether the signal comes in with certain signal strength x or a bigger signal strength x+1 makes no difference for the movements of the ship. It is the informational content as such that is causally relevant here, not the strength of the signal. In fact it can be causally strong, even if the signal is weak. This seems indeed like a primitive form of mental causation. Mental causation means that mental content qua content has causal efficacy.

If nature has – even at the quantum level – the capacity to represent mental content and act on the mental content as such, then there is a form of intentionality built into nature. If account (4) above is correct, then Bohm's theory implies a form of panpsychism in which some form of consciousness is ubiquitous in nature. If account (1) is correct, then there is some form of unconscious intentionality even at the very basic levels of nature. Fully developed consciousness arises from the complex configuration of these basic forms of consciousness. In any case, that Bohm places intentionality into the heart of matter via his theory of active information corroborates the thesis that the concept of intentionality links quantum mechanics to the philosophy of mind.

7 Many-worlds and many-minds theories and intentionality

Denial of (III): It is not the case that a measurement yields a unique result at the expense of other nomologically possible results. There is branching of the universe. All results are realized.

The many-worlds and the many-minds interpretation of quantum mechanics stems from the denial of (III). There are no unique results of a measurement because all possible results are realized, with each being relative with respect to a world or to a mind. These theories deny in total the collapse of the wave function. The many-worlds interpretation implies that all possible histories are real, and each one of them represents a "world" or a universe (Everett 1956). The concept of 'decoherence' is used to single out observable classical worlds (ordering of the phase angles) in a quantum superposition. This forking of the world into ever more worlds can be understood without making any reference to minds. In the same way collapse theories can construe the collapse both with or without reference to minds, the many-world theorist need not invoke minds to explain the forking of the world. In this case there is no real connection between this interpretation of quantum mechanics and the philosophy of mind.

But one can also see the forking as a multiplicity of different subject-object states, wherein for each branch there is a mind observing it. As this isolated mind does not know about the other minds observing the other branches, it will be puzzled as to why exactly this particular result (the result this mind observed) occurred. The mind might speculate about a mysterious collapse of the wave function, but in fact there was no collapse. The process of decoherence realizes all possible future states relative to a mind. This many-minds interpretation is a variant of the many-world interpretation and was first proposed by H.D. Zeh (1970). Later David Albert and Barry Loewer (1988) developed an influential version of it. Zeh's original idea was quite straightforward, and his goal was to avoid distinct worlds in a decohering universe without necessitating collapses of the wave functions. His idea was that there is a psycho-physical parallelism between decoherent physical states and minds. The metaphysical nature of this parallelism can be spelled out in different ways. It could be that the minds supervene on the physical states, it could also be the case that—in a more interactionist-dualist fashion—the minds actively select the physical states they correspond to. In any case the theory postulates a vast number of minds. No minds are split, they existed all along, far more than the individual minds of human persons in the common sense world. In this many-minds interpretation it is obvious that decoherence exists only in relation to a mind which represents its environment from a certain point of view. Representation by a mind, however, entails some form of intentionality. Here again it is intentionality that connects this interpretation of quantum mechanics to the philosophy of mind. The role consciousness plays in this theory depends—as might be expected by now—on how we construe the relationship between intentionality and consciousness. If we follow account (1), then the minds of the many-minds theory need to be conscious. They do feature intentional, representational states, but these states might be in many cases not complex enough to be classified as conscious because phenomenal consciousness requires some form of higher-order meta-representation. If we follow account (4), then there is a strong modal tie between intentionality and consciousness, and in any of the many minds that represents a decoherent physical state there exists a phenomenal consciousness of some sort. In fact, directly following from this, there will be a vast number of conscious minds in the universe, many more than our common sense view takes for granted.

8 Taking stock

We started out with the observation that the philosophical interpretation of quantum mechanics is far from being a closed issue. The many conflicting theories are, for the time being, consistent with the data and the mathematical calculations, and philosophers of mind tend to cherry-pick the interpretation of quantum mechanics that best fits their philosophical preconceptions. As a result, quantum mechanics cannot provide a compelling case for some possible solution to the mind problem at the expense of rival philosophical theories. In fact, some interpretations of quantum mechanics—like the GRW collapse theory and the many-worlds theory—do not seem to have substantial repercussions in philosophy of mind. Many other interpretations of quantum mechanics are, however, directly relevant for and connected to the philosophy of mind. I have argued above that it is the idea of represented mental content, the idea of aboutness or intentionality that plays the crucial role here. In each of the three main strategies of developing an ontology for the quantum realm there are well-established theories that make heavy use of the idea of mental representation or some (proto-) form of intentionality. The role consciousness plays in these theories is much less clear. I have argued that the nature and scope of consciousness in these theories is (partly) determined by how their proponents see the relationship between intentionality and consciousness: if consciousness is derived from intentionality it will play a less fundamental role and if there is a strong modal tie between intentionality and consciousness, then the phenomenal mind will be present wherever there is intentionality, if only in a simple form. In any case, it is primarily the concept of intentionality that connects many (but not all) interpretations of quantum mechanics to the philosophy of mind.

References

Albert, D. 1994, "Bohm's Alternative to Quantum Mechanics", *Scientific American* (5): 58–67.
Albert, D., and Loewer, B. 1988, "Interpreting the Many-worlds Interpretation", *Synthese* 77 (2): 195–213.
Bohm, D. 1952, "A Suggested Interpretation of the Quantum Theory in Terms of 'Hidden Variables'", *Physical Review* 85: 166–193.
Bohm, D. 1990, "A new theory of the relationship of mind and matter", *Philosophical Psychology* 3 (2): 271–286.
Bohm, D., and Hiley, B. 1993, *The Undivided Universe. An Ontological Interpretation of Quantum Theory*, London: Routledge.
Epperson, M. 2004, *Quantum Mechanics and The Philosophy of Alfred North Whitehead*, New York: Fordham University Press.
Everett, H. 1956, *Theory of the Universal Wavefunction*, Thesis, Princeton: Princeton University.
Ghirardi, G.C., Rimini, A., and Weber, T. 1985, "A Model for a Unified Quantum Description of Macroscopic and Microscopic Systems", in L. Accardi et al. (eds.), *Quantum Probability and Applications*, Berlin: Springer, 223 –232.
Goldstein, S. 2013, "Bohmian Mechanics", in E.N. Zalta (ed.), *The Stanford Encyclopedia of Philosophy* (Spring 2013 Edition), URL = <http://plato.stanford.edu/archives/spr2013/entries/qm-bohm/>.
Holt, J. 2003, *Blindsight & The Nature of Consciousness*, New York: Broadview Press.
Siewert, C. 1998, *The Significance of Consciousness*. Princeton: Princeton University Press.
Penrose, R., and Hameroff, S. 2011, "Consciousness in the Universe: Neuroscience, Quantum Space-Time Geometry and Orch OR Theory", *Journal of Cosmology* 14.
Stapp, H.P. 1989, "Quantum nonlocality and the description of nature", in J.T. Cushing and E. McMullin (eds.), *Philosophical Consequences of Quantum Theory*, Notre Dame, IN: University of Notre Dame Press, 154–74.
Stapp, H.P., 2007, *Mindful Universe. Quantum Mechanics and the Participating Observer*, Berlin: Springer.
Zeh, H.D. 1970, "On the interpretation of measurement in quantum theory", *Foundations of Physics* 1 (1), 69–76.

Antonella Corradini
Quantum Physics and the Fundamentality of the Mental

The question whether the mental is, or is not, a fundamental dimension of reality opens up a major divide among the perspectives currently held in the philosophy of mind. Thus I shall first examine the issue of fundamentality within the philosophy of mind and then turn to the analysis of the role that the fundamentality of the mental plays in quantum physics.

1 Fundamentality in the philosophy of mind

In *The Conscious Mind* David Chalmers affirms that,

> fundamental features cannot be explained in terms of more basic features, and fundamental laws cannot be explained in terms of more basic laws; they must simply be taken as primitive [...] The fact that consciousness does not supervene on the physical features shows us that (this) physical theory is not *quite* a theory of everything. To bring consciousness within the scope of a fundamental theory, we need to introduce *new* fundamental properties and laws" (Chalmers 1996, 126).

Theories which do not acknowledge the fundamentality of the mental are thus those which admit the supervenience of the mental upon the physical, such as nonreductive physicalism or emergentism, in any of its versions most similar to nonreductive physicalism.

As regards the theories which champion the fundamentality of the mental, I shall quote David Chalmers once again:

> I think that substance dualism (in its epiphenomenalist and interactionist forms) and Russellian monism (in its panpsychist and panprotopsychist forms) are the two serious contenders in the metaphysics of consciousness, at least once one has given up on standard physicalism. (I divide my own credence fairly equally between them.) (Chalmers 2013, forthcoming).

So, for Chalmers substance dualism and panpsychism (in the form of Russellian monism) are the most plausible views that acknowledge the fundamentality of the mental. Now, Chalmers' partial intellectual devotion to substance dualism surprises me for two reasons. The first one is that in his (1996) book Chalmers' naturalistic dualism assumes the form of a property dualism that is pretty unlikely to be upgraded to a more radical kind of dualism (he notes, "the issue of what it would take to constitute a dualism of substances seems quite unclear to me", 125). As to the second reason, it is quite difficult to understand how a philosopher can have equal predilection for two conceptions—substance dualism and panpsychism—that, apart from sharing the claim of the fundamentality of the mental, have very little in common. In this part of my talk I shall try to show that dualism and panpsychism are not equally plausible candidates for explaining the fundamentality of the mental and to ask the following question: mind is part of the fundamental structure of the world, but what shape should an account of its fundamentality take, dualism or panpsychism?

As part of an attempt to ask this question, I shall examine the argument in favour of panpsychism that has been formulated by Thomas Nagel (1979) and worked out again by William Seager (1991) and Galen Strawson (2006).

As Nagel writes, panpsychism seems to follow from these four premises:
1. Material composition;
2. Nonreductionism;
3. Realism;
4. Nonemergence.

I shall now analyse these premises, starting from the second premise, that is to say nonreductionism.

Nonreductionism: "Ordinary mental states [...] are not physical properties of the organism [...] and they are not implied by physical properties alone" (1979, 181).

At first glance it might appear that the nonreductionism illustrated in the second premise could be understood as a kind of property dualism. In order to obtain a kind of property dualism one has to claim, as Chalmers puts it, that

> there are properties of individuals in this world—the phenomenal properties—that are *ontologically* independent of physical properties" (1996, 125; my italics).

On Nagel's construal are then mental properties ontologically independent from physical properties? My answer is in the negative. In fact, evidence against nonreductionism as property dualism is contained in the first premise of Nagel's argument, which concerns the material constitution of objects: "No constituents besides matter are needed", says Nagel at page 181 of *Mortal Questions*. Strawson also affirms that "real physicalism [...] must accept that experiential phenomena *are* physical phenomena" (2006, 4; my italics).

A first consideration to draw from these quotations then is that nonreductionism in Nagel's argument should not be conceived of as a kind of property dualism but, rather, as a variety of nonreductive materialism or as a *broad physi-*

calism, to use Chalmers' words. Let us say something about Chalmers' ideas. This author, in the half of himself in which he is a supporter of panpsychism, chooses a specific kind of panpsychism that he calls *constitutive Russellian panpsychism*. This is a *constitutive* panpsychism because "macroexperience is (wholly or partially) metaphysically grounded in microexperience" (2013, 7). It is further a *Russellian* panpsychism because it relies on the assumption that "physics reveals the relational structure of matter but not its intrinsic nature" (8). The intrinsic nature of matter, instead, is constituted by *quiddities*, which are, at least some of them, microphenomenal properties underlying the microphysical structure and playing fundamental microphysical roles like mass (8). Constitutive Russellian panpsychism maintains that microphenomenal properties both serve as quiddities and as the grounds for macrophenomenal properties (9).

Now, when Chalmers poses the question whether quiddities are physical properties, in order to answer this question he introduces the distinction between narrowly physical properties and broadly physical properties. The former only include the structural properties of microphysical entities, but exclude quiddities; the latter, instead, include both structural properties and quiddities. Thus these are not narrowly but broadly physical. Once this distinction has been made, Chalmers believes that the question whether quiddities are physical properties "becomes something of a verbal question" (10). The same holds for constitutive Russellian panpsychism, which is incompatible with narrow physicalism, but is a kind of broad physicalism. Once again, any dispute whether narrow or broad physicalism is really physicalism turns out to be purely verbal (10).

However, it is not obvious that everything turns out to be a verbal dispute, as Chalmers maintains. To be sure, it is also a verbal dispute how one prefers to define physicalism. Nevertheless, it is a matter of fact that quiddities are deeply rooted in the microphysical network and that they have to follow the dynamics—of aggregation and disintegration—of the physical particles to which they inhere. As we will see, if it weren't so, if the first premise of Nagel's argument about material composition didn't hold any longer, not even an essential feature of panpsychism such as the ubiquity of the mental would hold any longer. The conclusion that should be drawn, therefore, is that quiddities *are* physical properties, even if they are described according to a broad or liberal kind of physicalism.

Let us now turn to the third premise of Nagel's argument, that is, realism.

Realism: In spite of nonreductionism, ordinary mental states "are properties of the organism, since there is no soul, and they are not properties of nothing at all" (1979, 182).

This premise is a direct consequence of the acceptance of materialism. As Strawson (2006) maintains, both physical and experiential phenomena are concrete and spatio-temporal, hence a non spatio-temporal substance to which mental states inhere cannot exist. This, however, is not a confutation of substance dualism, but a claim that derives from a previous assumption that materialism is true. But materialism may cause unwanted troubles, as can be seen from the following objection that Nagel, the author of "What is it like to be a bat", raises against himself:

> For Realism as I have defined it to be true, physical organisms must have subjective properties. What seems unacceptable about this is that the organism does not have a point of view: the person or creature does. It seems absurd to try to discover the basis of the point of view of the person in an atomistic breakdown of the organism, because that object is not a possible subject for the point of view to which the person's experiences appear (Nagel 1979, 189).

The problem—Nagel goes on—is that there is no alternative to this position: "I assume that neither I nor the mouse has a soul, to bear these mental properties" (189). But, once again, this assumption of the non-existence of a soul is justified only in virtue of the materialist background that Nagel endorses.

A final source of criticism arises from the sequel of the previous sentence:

> And even if we did, it would not remove the problem, because insofar as it is possible to grasp the idea of a nonmaterial *thing* (my italics), there is just much difficulty in understanding how *it* could have a point of view. But if the occurrence of a subjective experience is not the possession of a property by *something*, what is it? (189-190).

Nagel does not have a clear idea of the difference between the material substance and the mental one. On his view both are ultimately similar, because both are a *something*. However, the mental substance actually coincides with the conscious self, and the soul can be seen as the element that allows the organism to become a conscious self. Thus, if we admit the existence of a mental substance, which Nagel cannot do, due to his adhesion to materialism, we obtain a subject of experience, endowed with a first person point of view and first person experiences. Thus Nagel cannot answer the criticism he makes to himself without abandoning materialism, which is, however, an essential element of his argument in favour of panpsychism. As we will see later, the topic of how to conceive the subject is of central relevance in the discussion between dualism and panpsychism.

We have found that Nagel's argument is developed in the context of nonreductive materialism. One of the most widespread ways to combine these two positions—materialism and nonreductionism—is emergentism. On this view, phenomenal properties are properties belonging to higher complexity levels that are not deducible and not explainable from physical properties. Broad (1925, 52), in particular, distinguishes between trans-physical and intra-physical emergent laws. Within the latter, the trans-ordinal laws connect phenomena belonging to different levels of complexity (for example H_2O molecules and the phenomenal quality of liquidity), while trans-physical laws establish the ultimate character of phenomenal properties: i.e. their non reducibility to physical data. Broad acknowledges a different epistemological status to the two kinds of laws. Trans-ordinal laws are empirical hypotheses, whose truth or falsity is discovered only after emergent phenomena have, or have not, taken place. Trans-physical laws, instead, are necessarily of an emergent kind. This means that a mathematical archangel who knew everything about the microstructure underlying the qualitative aspects of the macrostructure would be unable to predict them (70). What is at issue here is not the limitations of our knowledge, but the existence of an unbridgeable explicative gap between the physical and the experiential level. This thesis is very close to contemporary arguments that intend to show the irreducibility of conscious experience to the underlying neurophysiologic phenomena (Jackson 1982, Levine 1983, Chalmers 1996). However, such a thesis does not enjoy the favour of those who argue for panpsychism, as Nagel's four premise, "nonemergence", shows.

Nonemergence: "There are no truly emergent properties of complex systems. All properties of a complex system that are not relations between it and something else derive from the properties of its constituents and their effects on each other when so combined. Emergence is an epistemological condition" (1979, 182).

What is at stake in the long controversy between emergentists and panpsychists? Let us follow the point of view of Strawson (2006). For Strawson it must be true by the definition of "emergent" that "any emergent phenomenon, say Y, is wholly dependent on that which it emerge from, say X". The total dependence of Y from X is the guarantee that Y is emergent from X (14). As we can see, his idea of emergence is very far from that of British Emergentists, who focussed not on the dependence, but on non-deducibility, non-explicability, non-predictability, in short on the novelty of emergent phenomena upon their physical bases. But towards this *brute* emergence, which ignores the relationships of dependence from the physical, Strawson's criticism is very sharp:

> Does this conception of emergence make sense? [...] I think that it is incoherent, in fact, and that this general way of talking of emergence has acquired an air of plausibility [...] for some simply because it has been appealed to many times in the face of a seeming mystery (12).

We can illustrate Strawson's position by recourse to the analogy he proposes between liquidity and phenomenal properties. In the case of liquidity, this is emergent upon water molecules because liquidity is entirely reduced to physical phenomena:

> We can see that the phenomenon of liquidity arises naturally out of, is wholly dependent on, phenomena that do not in themselves involve liquidity at all" (13–14).

In the emergentists' language, the case of liquidity falls under the trans-ordinal intra-physical emergent laws, whose epistemological status is that of an empirical hypothesis. In the case of phenomenal properties, instead, the analogy with liquidity does not hold, on Strawson's view, because it is absurd to think that when non-experiential phenomena stand in certain relations they *ipso facto* constitute experiential phenomena (5–16). This case would be covered, in the emergentist context, by trans-physical laws, which establish the a priori necessary irreducibility of the experiential properties to the physical ones. But for Strawson, as already said, this is *brute* emergence, and

> Emergence can't be brute. [...] One problem is that brute emergence is by definition a miracle every time it occurs, for it is true by hypothesis that in brute emergence there is absolutely nothing about X, the emerged-from, in virtue of which Y, the emerger, emerges from (18).

Strawson thus maintains that the physical basis alone is not sufficient to generate emergent properties, if these are phenomenal. How to find then a remedy to the lack of a sufficient condition for the emergence of the experiential? Given Strawson's conservative conception of emergence, he can affirm what follows:

> For what we do, when we give a satisfactory account of how liquidity emerges from non-liquidity, is show that there aren't really any new properties at all. Carrying this over to the experiential case, we get the claim that what happens, when experientiality emerges from non-experientiality, is that there aren't really any new properties involved at all. This, however, means that there were experiential properties all along; which is, precisely, the present claim (23).

This conclusion is very similar to that of the author who originally designed the argument for panpsychism. Nagel, in fact, concludes his argument by saying that

> If the mental properties of an organism are not implied by any physical properties but must derive from properties of the organism's constituents, then those constituents must have nonphysical properties [...] Since any matter can compose an organism, all matter must have these properties (1979, 182).

Strawson's choice in favour of panpsychism notwithstanding, he is aware that the intuition lying at the basis of panpsychism—that a mental phenomenon cannot be obtained from physical phenomena alone—induces others to support positions as different as dualism and eliminativism. I would like to ask whether there is an alternative response, of a dualistic kind, to the problem raised by Strawson, for which he found a solution in panpsychism.

If we agree with Strawson that at the microphysical level there must be something experiential, on a dualistic picture we could interpret this element as a peculiar, non material dimension of reality. Such a dimension has non material features because it grounds the possibility of the organism's development in a self, and this possibility would not be otherwise granted given the original diversity between physical and mental properties. On the other hand, this possibility is not something extrinsic to the organism but is rooted in this and in its potential capacities. Thus this possibility is a *power* in the genuine sense of the word, which is characterized by almost two properties that Molnar (2003, 60–81) attributes to powers: directedness and intrinsicality. As to the first feature, the organism's development is orientated from the very beginning towards the generation of a self. This cannot take place in the form of an unexplainable jump (it would be brute emergence) but through the manifestation, in the course of the organism's natural evolution, of gradually more complex properties and powers. As to the second feature, intrinsicality, it is crucial that the realisation of the original power (to become a self, to make emerge a self) does not take place in virtue of extrinsic relationships, but of properties which are intrinsic to the organism itself (102–108). On the interpretation that I here defend the organism's evolution towards the manifestation of the self does not lie on an occasionalist analysis of the connections between the mental and the physical. If things were this way, the psycho-physical nexus would depend on the intervention from factors external to the organism itself or it would be reduced to mere correlations. From the intrinsicality of the powers inherent to the organism it follows the pivotal idea of the mind-body co-evolution, on whose basis the realisation of non biological potentialities is induced by the development of the biological structure, which in its turn is afterwards affected by the causal activity of the conscious mind.

We thus reach a crucial point for the differentiation between dualism and panpsychism. It is worth noting that the process of actualisation of the self also implies its particularisation, its being the mind of a specific human being, or, for that reason, of a specific non human being. As we have just seen, the actualisation of the mind is induced by a biological process of high complexity, but increasing complexity is also a sign of increasing individualisation. Therefore the "pairing problem" (Kim 2001) is a pseudo-problem, given the intimacy which characterises the relationship between mind and body. So much said, we can now appreciate the fundamental difference between dualism and panpsychism, which regards the different ways in which they conceive of the subject. Panpsychism, in fact, grounds on a principle that is the opposite to individualisation, that is, on the ubiquitous presence of the experiential in all real entities. From the most elementary to the most complex levels there are different forms of experientiality. Common to all of them, however, is subjectivity, which makes all entities having experiential properties "subjects of experience". However, if we take into consideration the relationship between macro and micro-level, it is apparent that a plurality of micro-subjects of experience cannot constitute one macro-subject of experience, and this in virtue of the principle of privacy. This is the *combination problem*, whose solution seems to be quite difficult. Despite the numerous attempts to find acceptable solutions to this problem, the error underlying it appears to be basic. It consists in extending to the subjective dimension relationships which are inherent to the objective dimension. This holds for the combination of micro-subjects to form a macro-subject but also for the relation of grounding (Chalmers 2013), which does not appear in the right place if applied to a subject, who is by definition a simple entity.

As a conclusion of this part, a last comment on Nagel's essay on panpsychism. While, as I have tried to show, I disagree with Nagel about his claim that each premise of the argument for panpsychism is more plausible that its denial, I agree with him that each premise is not more plausible than the denial of panpsychism (1979, 181).

2 Fundamentality of the mental and quantum physics

One of the main philosophical distinctions between classical physics and quantum physics is that for the former physical properties are definable independently from the subject. For the latter, instead, precisely the opposite holds, that is to say physical properties are not definable without the contribution of the subject. The distinction between primary and secondary qualities can help us to clarify this point. According to classical physics the world image describes the primary properties of physical objects. These are properties intrinsic to physical objects, thus they belong to them independently from being observed by the subject. Secondary properties have instead a dispositional character, since they are those properties which appear to belong to the physical object only if the subject observes them. For example, X is red iff it appears as red every time it is observed. In Galileian science, the requirement of objectivity, conceived of as independence from the subject, amounts to the claim that primary properties alone are scientifically allowable. As is known, in the context of microphysical processes all observations have the character of an intervention which modifies the reality to be observed. From this it follows that properties of objects or of physical systems can be only understood as dispositions to yield certain results, on the basis of observations and with a certain probabilistic measure. This situation does not allow us to establish a clear-cut distinction between intrinsic primary properties and dispositional secondary properties any longer. In other words, in the realm of quantum physics all empirical properties have a secondary character. Therefore, as secondary properties are dispositional properties as regards the subject and thus presuppose its existence, all empirical properties characteristic of the quantum language are dispositional properties as regards the observer and thus presuppose its existence. As to classical physics, however, secondary properties are excluded from its language, which implies the non necessity of the subject for the description of reality. The opposite holds for a quantum description, for which a subjective component is essential.

Somebody could object that such a subjective component is eliminable. The supporters of this thesis maintain that an observing apparatus having a wholly physical structure can play the role attributed to the subject, so that a quantum observation is not a kind of conscious, first person knowledge but a simple variation of macroscopic physical bodies. However, on the subjectivist interpretation of quantum physics the role of the subject is indispensable, because nature itself does not make the distinction between observed object and measuring device. Such a distinction is essential to confer to the pointers of the measuring instruments the meaning that they have regarding the micro-physical reality to be observed. Thus every description of a quantum system is a description for an observer. A quotation from von Neumann (1932) illustrates this point:

> we must always divide the world into two parts, the one being the observed system, the other the observer. [...] Indeed experience only makes statements of this type: an observer has made a certain (subjective) observation; and never any like this: a physical quantity has a certain value (1932, 420).

On the same line of thinking, London and Bauer write that

> a coupling (between an object and a measuring device) is not yet a measurement. A measurement is achieved only when the position of the pointer has been *observed*. [...] We note the essential role played by the *consciousness* of the observer in this transition from the mixture to the pure case. Without his effective intervention, we would never obtain a new ψ function" (1939, 251).

If we now reflect upon the features displayed by the observing subject, we realise that she performs a knowledge act, that is, a conscious, first person knowledge act. But the observing subject is not only a knowing subject, she is also an agent, because it is the subject's free will which decides when and how to perform the observation (this point has been emphasized in particular by Henry Stapp in this volume). Thus the subject has not just a theoretical dimension but also a practical one. Returning briefly to the previous discussion on panpsychism, the conception of the subject that derives from quantum physics is much richer than that of a phenomenal self, that is mostly the focus of the debates on panpsychism. As we have just seen, it is the conception of a self who is capable of conscious, first-person knowledge, and of agency, based on her free intentionality. All these features of the subject cause difficulties for panpsychism, which is already in trouble, as we have seen, with the mere phenomenal aspects of the subject.

Now I would like to pause for a while to reflect on the philosophical consequences that should be drawn from quantum physics and its conception of the subject. The first remark is that the subjectivity implied by quantum physics is a good point in favour of those who champion the idea of the fundamentality of the mental. But, just as in the philosophy of mind the fundamentality of the mental can be conceived of in different ways, the same holds for the fundamentality of the mental implied by quantum physics. The second question is then whether we are entitled to say that in the context of quantum physics also dualism is the most convincing conception of the relationship between the mental and the physical.

As a matter of fact, a form of dualism has been developed, that is to say *polar dualism*, which is inspired by the findings of quantum physics (von Kutschera 2006, 2009). It is a kind of dualism characterised by the following principles:

1. Reality is constituted by two dimensions, the psychical and the physical, each of which must be considered as a non eliminable part, as a part not reducible to the other.
2. The psychical is related to the physical side of reality not as object to object, but as subject to object. In this way the mutual irreducibility of both aspects is saved. On the one hand, the physical pole is the intentional object of the psychical dimension (without object no subject may occur). On the other, the psychical dimension must be considered as a non eliminable part of reality. This is not reducible to the material object because the image of material reality cannot be constituted without an essential reference to the first person perspective that is characteristic of the subject (without a subject the physical image of reality does not make sense). The following quotation from von Kutschera underlines the importance both of the subject and the natural processes in the constitution of the intentional domain of physical knowledge:

   Physics presupposes observers, thus not all observers can be the product of physical evolution, as materialism maintains. Without subjects there can be only those processes that quantum physics describes without the projection postulate, that is to say those systems in which the probabilities of observational results develop over time without external interventions" (2006, 217; my translation).

3. Psychical reality is originally correlated with physical reality and the other way round, therefore it is not necessary to explain why they are correlated. On the contrary, if the two realities were juxtaposed to one another, it would be necessary to explain the reason of their interaction.

Clearly, the conception of polar dualism developed by von Kutschera is influenced by the subjectivist interpretation of quantum physics originating from von Neumann's work on the foundations of quantum theory. However, there are other philosophical approaches that also rely on quantum physics but begin with different assumptions. Even without disavowing the irreducibility of the psychical to the purely material dimension of reality, they conceive of the relationship between the mental and the physical as two aspects of a single reality, neither mental nor physical. On these dual-aspect theories the mental and the material dimensions are conceived of on the one hand as mutually irreducible, and on the other as reducible to an underlying dimension neutral with respect to both of them. A typical view of this kind is the Pauli-Jung theory, that understands the relationship between the mental and the physical as a relationship between the conscious dimension of the psychical and the dimension of local realism of classical physics. However, these dual aspects are formally implied by the psycho-physical basis, in which both the "holistic realism" typical of quantum reality and the "collective unconscious" typical of the original psychical reality are present (on this see Atmanspacher 2011, 20). Conceptions of this kind are not the result of a mere reflection upon the foundations of quantum physics, but are much more demanding since they propose real global ontologies grounded on the physical theory that nowadays captures most closely the fundamental aspects of reality, that is to say quantum physics. It is just their status as global ontologies that make them interesting for the topic of this volume but at the same time exposes them to forms of typically ontological criticism. Two of their aspects are worth noting in this regard: the nature of the neutral basis and the mode of causation of the dual aspects from this basis. Now, what is meant for neutral basis? Two main meanings are possible. On the first one for neutral basis we mean a reality that has got two fundamental aspects from the very beginning. At the onset these aspects are only potential but with time and thanks to evolution take gradually more complex and differentiated forms. In this case, a conception that appeals to that basis as a psycho-physical origin of the current differentiated reality would make sense. On the second meaning, however, the neutral basis is understood as undifferentiated, and this raises difficulties. On the one hand, in fact, it is not easy to understand what the neutral basis actually is and, on the other, it is quite incomprehensible why the dual psycho-physical reality should emerge from this basis. How should it emerge, in fact? First of all it does not seem that the derivation of dual reality from the neutral basis can be of an epistemic kind. Indeed, it could not be the result of the fact that neutral psycho-physical reality would be likely to appear according to the psychical and physical aspects. In order to appear, in fact, it should appear to a subject and a subject could not exist before the emergence of the subjective perspective. It does not make any sense to speak of the appearing of two different dimensions if this is not an appearing to a subject.

The derivation could therefore be of an ontological nature. This means that neutral reality should be its cause. But how is it possible that the undifferentiated neutral basis could be the cause of the differentiated dual aspects? It should be a formal cause that, given the neutral nature of the basis, could not originate from it. This conclusion can be naturally extended to those panpsychist conceptions that are inclined to understand the presence of psychical and physical properties in the fundamental particles as two aspects emerging from a neutral basis.

In conclusion, a polar kind of psycho-physical dualism seems to be the conception that is most coherent with the image that quantum physics currently proposes of reality. Dualism safeguards the irreducibility of both aspects and at the same time the co-essentiality of both of them. Without the subject it could not be possible to account for the secondary nature of all empirical properties. On the other hand the subject, as an observer, could not do without the physical world, inasmuch as this is the essential subject of her intentional perspective.

References

Atmanspacher, A. 2011, "Quantum Approaches to Consciousness", in E.N. Zalta (ed.) *The Stanford Encyclopedia of Philosophy* (Summer 2011 Edition); http://plato.stanford.edu/archives/sum2011/entries/qt-consciousness/, accessed on April 15, 2013.

Broad, C.D. 1925, *The Mind and its Place in Nature*, London: Routledge and Kegan Paul.

Chalmers, D. 1996, *The Conscious Mind*, New York-Oxford: Oxford University Press.

Chalmers, D. 2013, "Panpsychism and Panprotopsychism", *Amherst Lecture in Philosophy*, forthcoming.

Jackson, F. 1982, "Epiphenomenal Qualia", *Philosophical Quarterly*, 32: 127–136.

Kim, J. 2001, "Lonely Souls: Causality and Substance Dualism", in K. Corcoran (ed.), *Soul, Body and Survival*, Ithaca and London: Cornell University Press, 30–43.

Kutschera, F. von 2006, *Die Wege des Idealismus*, Paderborn: Mentis.

Kutschera, F. von 2009, *Philosophie des Geistes*, Paderborn: Mentis.

Levine, J. 1983, "Materialism and Qualia: The Explanatory Gap", *Pacific Philosophical Quarterly* 64, 354–361.

London F., and Bauer, E. 1939, *La theorie de l'observation en mécanique quantique*, Paris: Hermann. English translation: "The theory of observation in quantum mechanics", in J.A. Wheeler and W.H. Zurek (eds.) 1983, *Quantum Theory and Measurement*, Princeton, New Jersey: Princeton University Press, 217–259.

Molnar, G. 2003, *Powers: A Study in Metaphysics*, ed. by S. Mumford, Oxford: Oxford University Press.

Nagel, T. 1979, "Panpsychism", in Id., *Mortal Questions*, Cambridge: Cambridge University Press, 181–195.

Neumann, J. von 1932, *Die mathematischen Grundlagen der Quantenmechanik*, Berlin: Springer. English translation: *Mathematical Foundations of Quantum Mechanics*, Princeton, New Jersey: Princeton University Press 1955.

Seager, W. 1991, *Metaphysics of Consciousness*, Routledge, London-New York.

Strawson, G. 2006, "Realistic Monism: Why Physicalism Entails Panpsychism", in D. Skrbina (ed.) *Mind that Abides. Panpsychism in the new millennium*, Amsterdam-Philadelphia: John Benjamins Publishing Company, 33–56.

Jeffrey A. Barrett
Quantum Mechanics and Dualism

1 Quantum measurement and the temptation of dualism

The quantum measurement problem is arguably the most difficult conceptual problem in the foundations of physics. It is an indication of its difficulty that attempts to solve it have led physicists and philosophers of physics to speculate concerning the relationship between physical and mental states. We will consider the sense in which this relationship provides a degree of freedom that is tempting to use in addressing the measurement problem. We will start with Eugene Wigner's understanding of the standard collapse formulation of quantum mechanics.

Two years prior to being awarded the Nobel Prize in Physics, Wigner published a paper arguing that a consistent formulation of quantum mechanics requires one to endorse a strong variety of mind-body dualism. In particular, he argued:

> Until not many years ago, the 'existence' of a mind or soul would have been passionately denied by most physical scientists. [...] There are [however] several reasons for the return, on the part of most physical scientists, to the Spirit of Descartes' 'Cogito ergo sum' [...] When the province of physical theory was extended to encompass microscopic phenomena, through the creation of quantum mechanics, the concept of consciousness came to the fore again: it was not possible to formulate the laws of quantum mechanics in a consistent way without reference to consciousness.

And continued:

> It may be premature to believe that the present philosophy of quantum mechanics will remain a permanent feature of future physical theories; it will remain remarkable, in whatever way our future concepts may develop, that the very study of the external world led to the conclusion that the content of the consciousness is an ultimate reality (1961, 168–169).

To see why Wigner believed that quantum mechanics requires a commitment to a strong variety of mind-body dualism for its consistent formulation, one must understand the basic structure of the standard von Neumann-Dirac formulation of quantum mechanics to which he was committed, and the quantum measurement problem.

We will start with the standard theory and the measurement problem, then consider Wigner's argument. We will then consider a fragment of an argument from an earlier letter from Wolfgang Pauli to Max Born. This line of argument will lead us to consider how even a no-collapse formulation of quantum mechanics may commit one to a strong physical-nonphysical dualism on plausible-sounding assumptions. The suggestion, however, will be that while it is tempting to commit to some form of dualism to address the measurement problem, there are viable options for avoiding a commitment to a strong mind-body dualism.

2 The standard formulation of quantum mechanics

The standard-Dirac collapse formulation of quantum mechanics is based on four rules. There are two representational rules (1) *representation of states*: the state of a physical system S is represented by a vector ψ_s of unit length, sometimes called the wave function, in a Hilbert space H and (2) *representation of observables*: every physical observable O is represented by a Hermitian operator $\hat{O}$ on H, and every Hermitian operator on H corresponds to some observable. An interpretational rule (3) *interpretation of states*: a system S has a determinate value for observable O if and only if the system is in an eigenstate of the observable $O\psi_s = \lambda\psi_s$. And two dynamical laws (4a) *deterministic linear dynamics*: if *no measurement* is made, the system S evolves in a deterministic linear way: $\psi(t_1)_s = \hat{U}(t_0,t_1)\psi(t0)_s$ and (4b) *random nonlinear collapse dynamics*: if a *measurement* is made, the system S randomly, instantaneously, and nonlinearly jumps to an eigenstate of the observable being measured, where the probability of jumping to φ_s when O is measured is $|\psi\varphi|^2$. The first dynamical law (4a) explains quantum interference effects, and the second (4b) ensures that measurements yield determinate outcomes and explains quantum probabilities.

The problem with this formulation of quantum mechanics is that while *measurement* occurs as an undefined primitive term in the theory, the two dynamical laws typically give different predictions for the post-interaction state of a measuring device and its object system depending on whether one considers the device to be a physical system like any other or a collapse-causing observer. More specifically, if one treats an observer as a physical system like any other, then one should use rule 4a for the interaction between the observer and her object system; but if one takes the observer to be somehow special and capable of causing collapses, then one should use rule 4b for the interaction. And, since the two rules typical predict different states, one gets a logical contradiction if one tries to apply both. Further, and of particular importance to Wigner, there are also empirical consequences for when each rule is taken to apply—a point central to his friend story, which we will consider in the next section. So the standard formulation of quantum mechanics is either (1) *logically inconsistent* if one thinks that observers and other measuring devices are physical systems like any other or (2) *incomplete* in an empirically significant way if one does not know how to identify systems that should count as measuring devices. This is the quantum measurement problem.

3 Wigner's proposal

Wigner's proposal for solving the measurement problem was simple:

> The important point is that the impression which one gains at an interaction may, and generally does, modify the probabilities with which one gains the various possible impressions at later interactions. In other words, the impression one gains at an interaction, called also the *result of an observation*, modifies the wave function of the system. [...] [I]t is *the entering of an impression into our consciousness which alters the wave function* because it modifies our appraisal of the probabilities for different impressions which we expect to receive in the future. It is at this point that the consciousness enters the theory unavoidably and unalterably (1961, 172–173).

Importantly, while one might be tempted to read parts of this passage epistemically, Wigner took the collapse that resulted from the entering of an impression into the observer's consciousness to be a real physical process. As the Wigner's friend story makes clear, he took there to be experiments one might perform, at least in principle, to determine what systems cause collapses. His solution to the measurement problem, then, was to stipulate, as a fundamental principle of quantum mechanics, that a real physical collapse of the state occurs whenever a conscious mind gains the impression of the measurement result.

There is, indeed, a sense in which Wigner's proposal immediately solves the measurement problem by sharpening rules 4a and 4b. The dynamical laws are now (4a') *deterministic linear dynamics*: if *no conscious mind apprehends its state*, the system S evolves in a deterministic linear way: $\psi(t_1)_S = \hat{U}(t_0, t_1)\psi(t_0)_S$ and (4b') *random nonlinear collapse dynamics*: if a *conscious mind apprehends its state*, the system S randomly, instantaneously, and nonlinearly jumps to an eigenstate of the observable being measured, where the probability of jumping to φ_S when O is measured is $|\psi\varphi|^2$. If there is a simple determinate matter of fact concerning whether and when an impression enters into a consciousness, these sharpened rules provide a consistent specification for the quantum dynamics.

Wigner believed that this move was "required" for the consistency of the standard collapse theory, and he considered it to be the "simplest way out" of the quantum measurement problem (180). And, again, he took his specification of when collapses occur to have physical and empirical consequences. Namely, the state collapses caused by minds affect the quantum-mechanical states of physical systems and hence objective, observable properties of the physical world.

Wigner illustrated this with his friend story. Wigner's friend F has a measuring device M and both are ready to measure the x-spin of a spin-1/2 system S. The system S begins in the state

(1) $$1/\sqrt{2}(|\uparrow_x\rangle_S + |\downarrow_x\rangle_S).$$

If we use the linear dynamics, rule 4a, and assuming ideal correlating interactions, after the measuring device M interacts with the object system S and after the F looks at the pointer on the M, the composite system $F+M+S$ will be in the state

(2) $$1/\sqrt{2}(|"\uparrow_x"\rangle_F |"\uparrow_x"\rangle_M |\uparrow_x\rangle_S + |"\downarrow_x"\rangle_F |"\downarrow_x"\rangle_M |\downarrow_x\rangle_S).$$

This state follows directly from the linearity of the dynamical law and the assumption that the interactions perfectly correlate the x-spin of S and F's measurement record. By rule 3, this is a state where F has no determinate measurement record at all—indeed, he is in an entangled state with M and S here and hence does not even have a proper quantum-mechanical state of his own.

But if we use the nonlinear collapse dynamics, rule 4b, for the interaction between M and S, or for the interaction between M and F, or for when F's mind apprehends the state, the composite system $F+M+S$ will either be in the state

(3) $$|"\uparrow_x"\rangle_F |"\uparrow_x"\rangle_M |\uparrow_x\rangle_S$$

or in the state

(4) $$|"\downarrow_x"\rangle_F |"\downarrow_x"\rangle_M |\downarrow_x\rangle_S,$$

each with equal probability 1/2. In contrast with state 2 each of these states describe F as having a determinate measurement result on the standard eigenvalue-eigenstate link 3. In the first of these states, F determinately records the result "$\uparrow_x$" and in the second he determinately records the result "$\downarrow_x$".

Wigner argued that the state of the composite system must be either state 3 or state 4. To begin, Wigner believed that were he to ask the friend what the result of his measurement was, then he would hear his friend say something perfectly determinate. Then, after having completed the whole experiment, if he asked his friend, "What did you feel about the result of your measurement before I ask you?", the friend would certainly reply, "I told you already, I got the result ["↑$_x$" or "↓$_x$"]" as the case may be. That is, the friend would report that the result of his measurement "was already decided in his mind" before Wigner asked him. He concludes this line of argument:

> If we accept this, we are driven to the conclusion that the proper wave function immediately after the interaction of friend and object was already either [state (3)] or [state (4)] and not the linear combination [state (2)]. [...] It follows that the beating with the consciousness must have a different role in quantum mechanics then inanimate measuring device (1961, 176–177).

While Wigner recognized that it is not logically inconsistent to deny that the friend is right in reporting that he already had a determinate measurement result before he was asked, Wigner took such an option to be unacceptable. He argued that to deny that the friend has the same sort of determinate experiences that we do and hence causes collapses of systems to determinate property states "is surely an unnatural attitude, approaching solipsism, and few people, in their hearts, will go along with it" (1961, 177–178). So it is when the friend apprehends the state, and not when Wigner asks him what his result was, that the composite system collapses to a state where the friend has a determinate and now accurate measurement record.

The precise sense in which such collapses involve a real physical process that produces in principle observable results was important for Wigner's argument. Consider an observable $\hat{A}$ of the composite system $F+M+S$ that has

(5) $$1/\sqrt{2}(|"\uparrow_x")_F|"\uparrow_x")_M|\uparrow_x\rangle_S + |"\downarrow_x")_F|"\downarrow_x")_M|\downarrow_x\rangle_S)$$

as an eigenstate with eigenvalue +1, and

(6) $$1/\sqrt{2}(|"\uparrow_x")_F|"\uparrow_x")_M|\uparrow_x\rangle_S - |"\downarrow_x")_F|"\downarrow_x")_M|\downarrow_x\rangle_S)$$

as an eigenstate with eigenvalue –1. An observation of $\hat{A}$ would yield the result +1 with probability 1 if the interactions between F, M, and S are linear, and it would yield the result +1 with probability 1/2 and the result –1 with probability 1/2 if F's measurement somehow caused a collapse and state 3 or state 4 obtains. So, while extraordinarily difficult to perform due to the complexity of the object system and the difficulty in controlling for decoherence effects, there are at least in principle experiments that would determine what systems cause collapses, and hence what systems should count as conscious if, as Wigner argued, conscious apprehension causes collapses.

For his part, Wigner took the fact that his proposal had empirical consequences to be a virtue. In particular, it provided one a way, at least in principle, to determine which entities in fact cause collapses of physical states. The thought is that one might then compare this to one's pre-theoretic sense of which entities are conscious to test the theory's novel empirical predictions.

That said, one might naturally wonder whether Wigner was right to believe that a solution to the quantum measurement problem requires one to endorse a strong variety of mind-body dualism. The short answer is that this depends on the background assumptions one finds plausible and on the explanatory demands one places on quantum mechanics. If one believes, with Wigner, that there are collapses of the quantum mechanical state and that there must be a principled distinction between one type of system that always evolves linearly and another, strictly disjoint, type of system that causes collapses, then one might be similarly tempted to endorse quantum mind-body dualism. But very different commitments from Wigner's can also push one toward a commitment to a strong variety of mind-body dualism in the context of quantum mechanics. In particular, some sort of strong dualism may be required on plausible-sounding background assumptions even if one opts for a no-collapse formulation of quantum mechanics.

4 An argument for no-collapse dualism

In March 1954 Wolfgang Pauli wrote Max Born from to explain Einstein's objections to quantum mechanics. Einstein and Born had been debating by post the conceptual foundations of quantum mechanics. During his visit at the Institute for Advanced Study in Princeton, Pauli had read the letters, discussed them with Einstein, and come to believe that Born had completely misunderstood Einstein's position. Pauli wrote Born that "[i]t seemed to me that you had erected some dummy Einstein for yourself, which you then knocked down with great pomp" (1954, 221). Contrary to the popular view, a view also held by Born, that Einstein objected to the statistical nature of quantum mechanics, Pauli explained that Einstein's essential worry was not *determinism* but *realism*. In particular, Pauli reported that Einstein was concerned with how one assigned determinate properties like position to a physical system and, in particular, what happened when one observes a macroscopic object that is initially in a superposition of being at different positions.

Regarding what happens on observation, Pauli agreed with Einstein that "it is not reasonable to invent a causal mechanism according to which 'looking' fixes the position" (1954, 222). This put both Einstein and Pauli at odds with the dynamical postulates of the standard collapse formulation of quantum mechanics and, more specifically, Wigner's later position. Pauli, however, disagreed with Einstein that "a macro-body must *always* have a quasi-sharply-defined position in the 'objective description of reality'". Since Einstein believed that there was no collapse of the state on observation, if a macro-body is to have a quasi-sharply-defined position, then the standard quantum description had to be incomplete since it typically fails to specify even an approximately determinate position. Pauli, in contrast, accepted the standard quantum-mechanical state as a complete description of the *physical* state of the system.[1]

Pauli explained to Born why he disagreed with Einstein by appealing to the uniformity of nature. He reported, "I believe it to be *untrue* that a macro-body *always* has a quasi-sharply-defined position, as I cannot see any fundamental difference between micro- and macro-bodies". In particular, Pauli took the linear dynamics, rule 4a, always to hold, even during an observation. But since he also held that the quantum state provided a complete physical description and that an observation typically provides an observer with a determinate experience, he concluded that the appearance of the collapse of a system to a definite position during an observation was "a 'creation' existing outside the laws of nature, even though it cannot be influenced by the observer. The natural laws only say something about the *statistics* of these acts of observation" (1954, 223).

In contrast to Wigner's view where minds are responsible for collapses, Pauli's letter to Born suggests a no-collapse formulation of quantum mechanics where the linear dynamics always correctly describes the time-evolution of the state of every physical system but where the determinate mental state of an observer only statistically supervenes on the observer's physical state.[2] It does not take much to get from this fragment of an argument to a full argument for a strong variety of physical-nonphysical dualism if one is committed to no collapse

---

[1] More specifically, as Pauli explained in his 1948 essay "Modern Examples of Background Physics", that the physical state provided by quantum mechanics does not specify the value of an outcome in an individual case "does not mean an incompleteness of quantum theory within physics [...] but an incompleteness of physics within the whole of life" (translated and quoted in Enz 2002, 424).

[2] See Atmanspacher, and Primas 2006 for an extended discussion of Pauli's views regarding the relationship between physical and mental states. While Pauli's assumptions support a strong physical-nonphysical dualism, for his part, he wanted to somehow reconcile the nonphysical experience of an observer with the physical world. As he put the goal in a 1952 essay, "[i]t would be most satisfactory if physis and psyche could be conceived as complementary aspects of the same reality" (translated and quoted in Atmanspacher, and Primas 2006, section 5.1).

of the quantum mechanical state and rule 4a always correcting describing the time-evolution of the quantum state.

In particular, the following assumptions are sufficient to commit one to a strong variety of dualism:

- *Assumption 1 (state completeness)*: The standard quantum-mechanical state provides a complete and accurate representation of the physical state.
- *Assumption 2 (no collapse)*: The linear dynamics, rule 4a provides a complete and accurate description of the evolution of the physical state for all systems at all times.
- *Assumption 3 (empirical consistency)*: If a system is initially in a superposition of states corresponding to different eigenvalues of the observable being measured, then it is possible for the measurement to yield a result corresponding to any of those eigenvalues.
- *Assumption 4 (no branching)*: The measurement interaction between an observer and a physical system typically yields a single determinate measurement result.

By assumptions 1 and 2, a typical measurement interaction yields a physical state where the observer records a superposition of mutually incompatible measurement results. However, by assumption 4, the observer nevertheless has a single determinate measurement result. By assumptions 1 and 3, the value of the measurement result cannot supervene on her *physical state*. So, insofar as it supervenes on anything, the observer's measurement result must supervene on her *nonphysical state*. And one is committed to a strong physical-nonphysical dualism.

In order to see more clearly how this argument works, consider the Wigner's friend story again. If the post-measurement state predicted by the linear dynamics, the state described by expression 2, is the observer's complete physical state, then the observer's complete physical state clearly does not determine the result of her measurement. Indeed, since the physical state here is perfectly symmetrical between the two possible results here, there is nothing in the state that could determine one or the other.[3] So if the observer has a single determinate measurement result, it must be determined by something nonphysical, presumably the observer's nonphysical state. And one is hence committed to a strong physical-nonphysical dualism.

---

[3] Note that even if the physical state were not perfectly symmetric, the physical state would not be sufficient to determine the single result of the measurement since, for the theory to be empirically adequate, each result associated with a positive amplitude must be statistically possible, so neither can be determined by the physical state that is predicted by the linear dynamics.

It is sometimes suggested that decoherence considerations might explain why there is a single determinate measurement record when the post-measurement state is one like (2). Note, however, that linear interactions with the environment will simply entangle more systems with the state of the compost system $F+M+S$. Hence such interactions will do nothing whatsoever to produce a physical state that describes a system with a single determinate measurement record. Rather, in order for the observer's complete state to describe a single determinate measurement record when such a post-measurement state obtains, one must add something to the physical description given by (2) that specifies the value of that record. On the assumption that (2) is the *complete physical state*, what one adds to get the observer's complete state all told will be a description of something nonphysical.

Given the four assumptions above, then, an observer's determinate measurement records must supervene on a nonphysical aspect of the observer's complete state. Further, one might argue, for quantum mechanics to be empirically adequate, this aspect of the complete state must also be something to which the observer has epistemic access. The most direct way to ensure this would be to stipulate that the value of an observer's measurement outcome is determined by the observer's mental state, then make this state determinate. On this line of argument, one again ends up committed to a strong variety of mind-body dualism, strong because since the determinate outcome of the observer's measurement fails to supervene on her physical state.

One might have thought that starting with a no-collapse view would prevent one from having to say when collapses occur, as Wigner was required to do, and hence allow one to avoid a commitment to quantum dualism. But this is one half-right. While one does not have the problem of saying when collapses occur, one does have the problem of saying how an observer can have a determinate measurement outcome without a collapse of the entangled superposition like (2) and providing something in the full state description on which the value of that outcome might supervene. The most direct way to get determinate records that are epistemically accessible is to add them as the experiential state of the observer, but if one one takes the quantum state to provide the observer's complete *physical* state, then one ends up committed to a strong mind-body dualism.[4]

Not only are the reason for the quantum dualisms different, there are also significant differences between the Wigner's type of dualism and the no-collapse

---

4 Note that even if the physical state were not perfectly symmetric, the physical state would not be sufficient to determine the single result of the measurement since, for the theory to be empirically adequate, each result associated with a positive amplitude must be statistically possible, so neither can be determined by the physical state that is predicted by the linear dynamics.

dualism just described. Perhaps most salient is that, while minds cause collapses on Wigner's view, in the no-collapse dualism described, minds are just there to explain determinate measurement outcomes—they are just something on which determinate outcomes might supervene, and, as such, they need never affect physical states. Indeed, since the evolution of the physical state on a no-collapse theory is always given by the linear dynamics, which depends only on the physical state, there is a clear sense in which mental states cannot cause physical events here. The minds are just along for the ride following their own auxiliary dynamics, a dynamics that will be contingent on the evolution of the physical state.[5]

5 Considering the assumptions

If one does not like the strong variety of physical-nonphysical dualism they entail, and there is much not to like, one must give up one of the assumptions that go into the argument of the last section. Let's consider their relative plausibility.

State completeness is a leading candidate for sacrifice. This is the assumption that the standard quantum-mechanical state provides a complete and accurate representation of the physical state. This can be thought of as the assumption that there are no hidden variables unaccounted for in the standard quantum state. It has a long and distinguished history in the development of quantum mechanics. Taking the standard quantum description to be incomplete, Einstein famously denied this assumption. He believed that standard state was incomplete because it failed to specify the values of the real physical quantities that determined of measurement outcomes. At the time, Einstein was very much in the minority in criticizing this assumption. But, as we have seen here, there can be a significant conceptual cost to assuming that the quantum state provides a complete description of the *physical* state–in particular, one might then end up committed to there also being a *nonphysical* state.[6]

Since the linear dynamics entails post-measurement states like 2 and since such states do not select a single measurement result, if one insists on state completeness, one must either give up that there is a determinate measurement

---

[5] Of course, for a complete no-collapse theory one must clearly specify the dynamics for the evolution of mental states. Albert, and Loewer (1988) provide a concrete example for how to do this. See Barrett 1999 for a discussion of this and other options.

[6] See Einstein, Podolski, and Rosen 1935 for his extended argument that the standard quantum-mechanical state is incomplete.

result[7] or give up that there is just one measurement result[8] or give up on the complete *physical* state being sufficient to determine the measurement outcome.[9] None of these options is particularly attractive, and the last commits one to a physical-nonphysical dualism. But if one is going to add something to the full state description, then one might deny state completeness and make it something physical, but something beyond the standard quantum state, that determines measurement outcomes.

Bohmian mechanics provides a concrete example for how to do precisely this. On Bohm's theory, particle positions are always determinate, so, insofar as physical measurement records are determined by particle positions, measurement results are determinate as well.[10] More specifically, in the context of the Wigner's friend story, the theory explains how the position of the particles that make up the pointer of the measuring device M end up associated with one or the other of the two possible measurement results represented in the state 2 and how this association provides the friend *F* with an effective measurement record that one can expect to be well-correlated with whatever actions *F* makes on the basis of the value of that record. It also explains why one can expect such records to satisfy the standard quantum statistics. This is a long story involving a number of subtleties along the way, but since we know how to tell it, we know at least one concrete way to give up the state completeness assumption.[11]

Giving up the assumption of state completeness by adopting Bohmian mechanics, however, exchanges a strong physical-nonphysical dualism for a strong physical-physical dualism where the evolution of the wave function is described by one dynamical law and the motion of particles by another and where the positions of the particles do not supervene on the standard quantum mechanical state.[12]

---

[7] This is the strategy pursued by the so-called bare theory where one seeks to explain the belief that there is an ordinary determinate measurement outcome as an illusion predicted by the theory. See Albert 1992 or Barrett 1999 for descriptions.

[8] This is the strategy of the many-worlds interpretation where one has a world with a different measurement outcome corresponding to each term in the final superposition 2 written in the determine record basis. See Barrett 1999.

[9] This is the strategy of the single-mind and many-minds formulation of quantum mechanics. See Albert, and Loewer 1988.

[10] Bohmian mechanics needs the assumption determinate measurement outcomes supervene on determinate particle positions in the theory. While this is a plausible assumption given typical hamiltonians of interaction, it is also easy to say how such an assumption might fail. See, for example, Albert 1992 discussion of John 1 and 2.

[11] To start, see Bohm 1952 and the discussion of Bohmian mechanics in Barrett 1999.

[12] This line of argument is perhaps particularly compelling against Bohmians who are also wave-function realists. See Ney, and Albert 2013 for recent discussions of the metaphysics of

Indeed, one might argue that the wave-function/particle-position dualism of Bohmian mechanics looks very like the mind-body dualism of Albert and Loewer's (1988) single-mind theory. On each of these theories, the quantum-mechanical state evolves linearly and the hidden variable that determines measurement outcomes, particle positions in Bohm's theory and mental states on Albert and Loewer's theory, obey an auxiliary dynamics and remain always determinate.

One might further argue that the strong physical-physical dualism of Bohm's theory has no virtues over a variety of strong-mind body dualism. But I do not think that is right. Rather, it seems to me that there is an important distinction to be made between the two types of theory regarding the sort of account of mental states each allows. In particular, while a strong mind-body dualism of the sort that we have been discussing simply precludes such an explanation, Bohm's theory allows one to continue to seek an explanation of mental states by considering how they might supervene on physical states.

Another candidate one might sacrifice to avoid quantum dualism is the *no-collapse* assumption. This is the assumption that the linear dynamics provides a complete and accurate description of the evolution of the physical state for all systems at all times. This assumption, of course, is violated by the standard collapse formulation of quantum mechanics. Indeed, it is precisely this that leads to the quantum measurement problem in the first place—if the standard theory did not have the two mutually incompatible dynamical laws, one would not face the embarrassment of having to say when each obtains. And, of course, it was in addressing the measurement problem that Wigner argued that a commitment to a strong variety of mind-body dualism is required. Hence, one does not automatically escape a commitment to quantum dualism by allowing for collapses.

That said, we do know how to allow for collapses of the state without committing to a physical-nonphysical dualism. Collapse formulations of quantum mechanics like GRW (1986) provide prescriptions for how and when collapses occur without in any way appealing to a physical-nonphysical distinction. The original version of GRW, for example, stipulates that, while each typically obeys the linear dynamics, every particle has a very small, but positive, probability per unit time of collapsing to a state close to an eigenstate of position. The effect of this stochastic term in the dynamics is that while microscopic objects involving few particle will likely behave linearly, macroscopic objects involving many particles whose positions are strongly correlated will likely have an approximately determinate center of mass and behave quasi-classically. There perhaps a sort of dualism at work here, but it is purely physical and involves only the dynamics.

Bohmian mechanics and varieties of wave function realism in particular.

That said, there are good reasons not to like collapse theories at all, and hence to keep the no-collapse assumption. To begin, there is strong empirical support for the linear dynamics insofar it has always made the right empirical predictions whenever we have been able to isolate and control a physical system well enough to test it. Further, since it predicts the instantaneous collapse of specially extended systems, the collapse dynamics, as it stands, is incompatible with relativistic constraints.[13]

Concerning the *empirical consistency* assumption, it is unclear, at least to me, how one might sacrifice this on any empirically adequate formulation of quantum mechanics. This is the assumption that if a system is initially in a superposition of states corresponding to different eigenvalues of the observable being measured, then it is possible for the measurement to yield a result corresponding to any of those eigenvalues. The thought is that this is simply required by our experience given the way that we assign quantum-mechanical states. Even in Bohmian mechanics, where one has a fully deterministic theory and particle position as the only observable non-contextual property, if a system is initially represented by an effective wave function corresponding to different eigenvalues of the (possibly contextual) observable being measured, then there is a positive epistemic probability of the measurement yielding the (possibly contextual) result corresponding to any of those eigenvalues. The upshot is that it is difficult to see how one could given this up and still have something that is recognizable as quantum mechanics. If one gives a concrete proposal for how to do it, then one might consider the potential costs and benefits of sacrificing it.

Finally, the no branching assumption holds that the measurement interaction between an observer and a physical system typically yields a single determinate measurement result.[14] While this may seem entirely uncontentious, this assumption is famously given up on at least some reconstructions of Everett's pure wave mechanics, theories like the many-worlds interpretation.[15] Giving it up, however, comes with significant costs. Particularly salient among these, if one allows for branching where a copy of the initial observer determinately gets a different measurement outcome on each branch, it is difficult to make sense of the standard quantum probabilities. Indeed, the probability of an observer

---

13 See Barrett 2014 for a recent discussion.
14 The *typically* here is just supposed to cover the chance that something goes wrong with the measurement like the pointer breaking during the measurement yielding a state where one piece points at one result and the other at a different result.
15 See Barrett, and Byrne 2012 for a description of Everett's own views, and Wallace 2012 and Saunders et al. 2010 for a recent discussion of the many-worlds interpretation.

getting each result is one insofar as one understands the observer as surviving the branching process at all, and this is not the statistical prediction one wants from quantum mechanics. Further, concerning the topic at hand, if one allows for branching, one avoids a strong physical-nonphysical dualism only to find oneself with a strong physical-physical pluralism of alternative branches, each with copies of the original observer.[16]

The upshot is that while one would face nontrivial costs giving up any of the assumptions that lead to physical-nonphysical dualism in the no-collapse argument, we know concretely how to give up at least three of the four explicit assumptions. That said, we have also seen that giving up one or more of these assumptions does not automatically prevent one from ending up committed to some variety of dualism by one's favored resolution to the measurement problem. I take strategic considerations regarding theory choice and metaphysical commitment here to be a matter of cost-benefit analysis given one's predictive and explanatory values. The interesting discussion regards the details of the expiatory tradeoffs involved the alternative options.

6 Discussion

On this view, the threat of a commitment to a strong variety of dualism in quantum mechanics ultimately results from competing explanatory demands. The linear dynamics is needed to explain interference effects. But it cannot, by itself, explain how a measurement interaction yields a single determinate measurement record. Hence, if one demands an explanation of determinate measurement records in terms of objective features of the world, then one must add something to the theory. It is this addition that threatens a commitment to some strong variety of dualism or metaphysical pluralism.[17]

---

[16] For discussions of the possible metaphysical commitments of such an approach see Saunders et al. 2010 and the conceptual introduction in Barrett, and Byrne 2012.

[17] Perhaps unsurprisingly, there are approaches to quantum mechanics where one does not make this sort of realist explanatory demand. On Richard Healey's (2012) pragmatist mechanics, for example, state attribution is not directly representational of the physical state of a system, and, hence, one does not require an account of determinate measurement records in terms of attributed states. And one might not worry much about the dynamics since a quantum state represents something more like an agent's epistemic state than the physical state of a system on such a view. There are, of course, significant explanatory costs giving up on direct physical description.

One might add the collapse dynamics to get determinate measurement records. But then one has a theory with two dynamical laws and one must clearly say when each obtains. And, as Wigner argued, given that one only wants, or needs, a system's state to collapse when it is observed, a natural way to accomplish this is to stipulate that conscious observers cause collapses by dint of being conscious. This provided Wigner with principled distinction between systems that cause collapses and those that do not, and the determinateness of the observer's mental state on this view is never threatened by physical superposition. And one ends up committed to a strong variety of mind-body dualism.

But, as we have seen, one can also find oneself committed to a strong variety of mind-body dualism if one takes the standard quantum-mechanical state to provide a complete physical description and denies that there are ever collapses of the quantum mechanical state. If the linear dynamics always obtains but a measurement interaction typically yields just a single measurement result, then, since that single outcome cannot typically be represented by the superposed physical state, it must be represented by a nonphysical state. And since the outcome is meant to explain the observer's experience, it must be a nonphysical state on which the observer's experience supervenes.

The point here is not that quantum mechanics requires a commitment to a strong variety of mind-body dualism. Rather, there remain a number of other options on the table. While quantum mechanics does push toward some variety of pluralism, it need not be a physical-nonphysical dualism. Bohmian mechanics illustrates how one might add something physical to the quantum state to provide something on which determinate measurement records might supervene in a no-collapse theory. One ends up on that account with a strong physical-physical dualism where one must specify both particle positions and the standard quantum state to characterize a physical system. And GRW-type spontaneous collapse theories illustrate how one might specify a single dynamical law that incorporates a sort of physical dualism in its sometimes linear and deterministic and other times nonlinear and stochastic dynamics. And in Bohmian mechanics and GRW, the sort of dualism involved is arguably much more modest than the sort of mind-body dualism required by Wigner's account or something like Albert and Loewer's single-mind theory. While mental states do not typically supervene on physical states in the latter theories, there is nothing in the structure of the former that would prevent this. Such formulations of quantum mechanics, then, exhibit the methodological virtue of not automatically precluding one from explaining mental states by describing how they supervene on physical states.

Whether a satisfactory resolution to the measurement problem should be taken to require some variety of mind-body dualism, physical-nonphysical dualism, or physical-physical dualism depends on the precise explanatory demands one places on quantum mechanics and on the background assumptions one finds plausible. My sense is that if a set of plausible-sounding assumptions commits one to a strong any sort of physical-nonphysical dualism where the nonphysical states do not supervene on the physical states, then one should sacrifice some of the plausible-sounding assumptions. The puzzle is what to sacrifice.

Given the options, I take the least objectionable to be either (1) sacrificing physical state completeness and adopting a hidden-variable theory like Bohmian mechanics hence opting for a physical-physical dualism or (2) sacrificing the requirement that the complete state determine a single measurement outcome and adopting something like Everett's pure wave mechanics. To be sure, each of these options comes with significant explanatory costs.[18] But quantum mechanics should be expected to requires one to sacrifice at least come of one's pre-theoretic intuitions.

---

18 See the conceptual introduction of Barrett, and Byrne 2012 for a discussion of the conceptual costs of taking pure wave mechanics seriously.

References

Albert, D.Z. 1992, *Quantum Mechanics and Experience*, Cambridge, MA: Harvard University Press.

Albert, D.Z., and Loewer, B. 1988, "Interpreting the Many Worlds Interpretation", *Synthese* 77: 195–213.

Atmanspacher, H., and Primas, H. 2006, "Pauli's Ideas on Mind and Matter in the Context of Contemporary Science", *Journal of Consciousness Studies* 13 (3): 5–50.

Barrett, J.A., and Byrne P. 2012, *The Everett Interpretation of Quantum Mechanics: Collected Works 1955-1980 with Commentary*, Princeton: Princeton University Press.

Barrett, J.A. 1995, "The Single-Mind and Many-Minds Formulations of Quantum Mechanics", *Erkenntnis* 42: 89–105.

Barrett, J.A. 1999, *The Quantum Mechanics of Minds and Worlds*, Oxford: Oxford University Press.

Barrett, J.A. 2006, "A Quantum-Mechanical Argument for Mind-Body Dualism", *Erkenntnis* 65 (1): 97–115.

Bohm, D. 1952, "A Suggested Interpretation of Quantum Theory in Terms of 'Hidden Variables'", Parts I and II, *Physical Review* 85: 166–179, 180–193.

Einstein, A., Podolsky, B., and Rosen N. 1935, "Can Quantum-Mechanical Description of Physical Reality Be Considered Complete?", *Physical Review* 47: 777–780.

Enz, C.P. 2002, *No Time to be Brief: a Scientific Biography of Wolfgang Pauli*, Oxford: Oxford University Press.

Ghirardi, G.C., Rimini, A., and Weber, T. 1986, "Unified Dynamics for Microscopic and Macroscopic Systems", *Physical Review*, D 34: 470.

Healey, R. 2012, "Quantum Theory: A Pragmatist Approach", *British Journal for the Philosophy of Science*, 63 (4): 729–771. doi:10.1093/bjps/axr054.

Ney, A., and Albert D.Z. 2013 (eds.), *The Wave Function: Essays on the Metaphysics of Quantum Mechanics*, New York: Oxford University Press.

Neumann, J. von 1932, *Die mathematischen Grundlagen der Quantenmechanik*, Berlin: Springer. English translation: *Mathematical Foundations of Quantum Mechanics*, Princeton, New Jersey: Princeton University Press 1955.

Pauli, W. 1954, "Letter to Born, 31 March 1954", in M. Born (ed.) 1971, *The Born-Einstein Letters*, London: Walker and Co.

Saunders, S., Barrett J., Kent A., and Wallace D. (eds.) 2010, *Many Worlds? Everett, Quantum Theory, and Reality*, Oxford: Oxford University Press.

Wallace, D. 2012, *The Emergent Multiverse: Quantum Theory according to the Everett Interpretation*, Oxford: Oxford University Press.

Wigner, E.P. 1961, "Remarks on the Mind–Body Question", originally published in I.J. Good (ed.) 1961, *The Scientist Speculates*, London: Heinemann, 284–302. Reprinted in J.A. Wheeler, and W.H. Zurek (eds.) 1983, Quantum Theory and Measurement, Princeton: Princeton University Press, 168–181.

Part II: Quantum Physics, Consciousness, Agency, and Free Will

Ulrich Mohrhoff
Consciousness in the quantum world: An Indian perspective

1 Introduction

Imagine that both Alice and Bob see a measurement pointer, and that this indicates a particular value. In the good old days of classical physics, they would have agreed that there is a real pointer "out there", for they would have been in possession of a theoretical model of the world "out there", and they would have had no problem situating the pointer in that model. The quantum-mechanical situation is strikingly different. The theory's irreducible empirical core is a probability calculus. Because this presupposes the events to which, and on the basis of which, it serves to assign probabilities, it cannot account for the existence of these events. If one nevertheless tries to perform the classical conjuring trick, which consists in the reification of calculational tools, one obtains a theoretical model that fails to accommodate the very events that the theory serves to correlate. The challenge posed by the measurement problem is to demonstrate the consistency of the theory's correlation laws with the *measurement*-independent reality of value-indicating events. Whether or not these events have a *mind*-independent reality is an entirely separate issue.

Making physical sense of the mathematical formalism of quantum mechanics calls for an interpretative principle, and I know of only one such principle that meets this challenge. In the first part of my paper I shall put this principle to work in two kinds of experimental situation, and I shall arrive at the conclusion that reality cannot be modelled from the bottom up. Quantum theory's explanatory arrow points in the opposite direction: from unity to multiplicity. The quantum world is not put together, it is *manifested*. It is the manifestation of an intrinsically undifferentiated Being. This Being, as I shall argue in the second part of my paper, holds the key to the mysterious commerce between mind and matter.

Part I

2 An interpretative principle

I agree with John Bell (1990) that "measurement" is a bad word. What is bad is the suggestion that quantum mechanics presupposes experimenters who build instruments for a *purpose*, or observers who *take cognizance of* outcomes. The correlata of the quantum-mechanical correlation laws are measurement outcomes only in the restricted sense that they make available information about the values of observables.

Quantum theory is built on two Lorentz-invariant calculational rules (Mohrhoff 2009a). The first rule is consistent with classical probability theory. The second rule instructs us to add the *amplitudes* of alternatives where classically we would add their *probabilities*. We use the first rule if there are events that indicate which alternative has occurred, or if correlations exist that make it possible to predict the outcome of a measurement designed to determine which alternative has occurred. We use the second rule if nothing indicates which alternative has occurred and no such correlations exist.

So what necessitates the use of the second rule? The following interpretative principle provides the answer: *Whenever quantum mechanics instructs us to add the amplitudes of alternatives rather than their probabilities, the distinctions we make between the alternatives correspond to nothing in the physical world. They cannot be objectified. They exist solely in our minds.*

3 Alternatives involving distinctions between regions of space

As promised in the introduction, I shall apply this principle to two kinds of alternatives: alternatives involving distinctions between regions of space and alternatives involving distinctions between things.

An example of the first kind is the well-known two-slit experiment with electrons (Feynman et al. 1965).[1] When we are required to use the second rule, the distinction between "the electron went through the left slit" and "the electron

---

[1] Chapter 1, *Quantum Behaviour*.

went through the right slit" cannot be objectified. All we can say is that it went through the union of the two slits. But if an electron can pass through the union of two regions of space without passing through either region, then the distinctions we make between parts of space cannot be intrinsic to space.

So what furnishes space with its so-called parts? The short answer is: detectors. By means of its measurement-independent macroscopic properties a detector realizes a "region of space", and thereby it makes it possible to attribute to a microscopic object the property of being in that region.

But if it is impossible to attribute to a physical object the property of being in a "region of space" unless this property is made available for attribution by a detector, then the spatial differentiation of the physical world cannot be complete—it cannot go "all the way down". Because the uncertainty principle rules out the existence of definite relative positions (except non-relativistically, in the unphysical limit of infinite momentum dispersion), detectors cannot realize definite positions, and this makes it impossible to attribute a definite position to anything else. We can therefore conceive of a partition of space into *finite* regions so small that none of them is available for attribution.

But if the spatial differentiation of the physical world is incomplete, then there are objects whose positions are indefinite only in relation to an imagined spatial background that is more differentiated than the physical world (Mohrhoff 2009b). If we reserve the adjective "macroscopic" for these objects and their positions, then every outcome of a measurement of a macroscopic position is *necessarily* consistent with both quantum mechanics *and* the laws of motion that quantum mechanics yields in the classical limit—to the extent that they are testable. (The only exception occurs when a macroscopic object—the proverbial pointer needle—indicates the value of an observable). Nothing therefore stands in the way of attributing measurement-independent reality to the positions of macroscopic objects.

4 Alternatives involving distinctions between things

What can we learn by applying the above interpretative principle to alternatives involving distinctions between things?

Consider four non-overlapping regions. Initial measurements indicate the presence of one particle in region A and one particle in region B. We wish to calculate the probability of finding one particle in region C and one particle in region D. There are two alternatives. In situations in which we are required to add amplitudes, the distinctions we make between the alternatives cannot be objectified. In these situations the particles are neither individuating substances nor do they carry individuating properties.

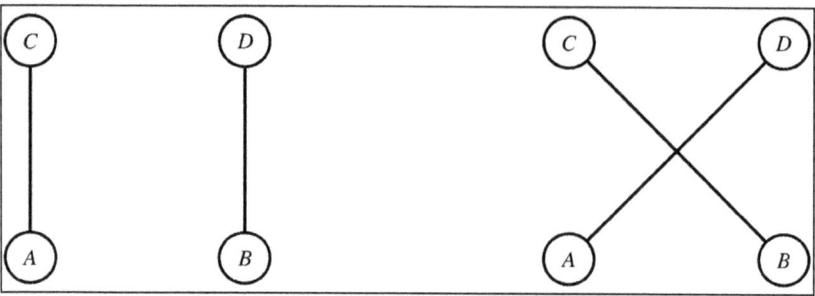

Quantum mechanics challenges us to think in ways that do not give rise to unanswerable questions. If we take for granted that space is an intrinsically differentiated expanse, we are led to ask the unanswerable question, "Through which slit did the electron go?". If we take for granted that initially there are *two* things, and that subsequently there are again *two* things, we are led to ask the unanswerable question, "Which incoming particle is identical with which outgoing particle?". On the other hand, if we adopt the view that initially there is *one* thing present in both region A and region B, and that subsequently there again is *one* thing present in both region C and region D, this unanswerable question can no longer be asked.

At any rate, nothing prevents me from taking the view that what is simultaneously present in two places, both initially and the next time we check, is one and the same thing. Nor need the numerical identity of what presents itself *here* and what presents itself *there* be confined to particles of the same type. There is no compelling reason to believe that this identity ceases just because it ceases to have observable consequences when persistent distinguishing characteristics exist. What can be present in different places can also be present with different properties other than position. Nothing therefore stands in the way of the view that *intrinsically* each fundamental particle is numerically identical with every other fundamental particle.

5 Manifestation

If this is the case, and if in addition the spatial differentiation of the physical world is incomplete, then reality cannot be modelled from the bottom up, whether on the basis of an intrinsically and completely differentiated space or spacetime or out of a multitude of separate building blocks. Quantum theory's explanatory arrow points in the opposite direction: from unity to multiplicity, from a single undifferentiated Being, which exists in an anterior relationship to spatial distinctions, to a world of forms— forms that resolve themselves into spatial relations between formless particles, relations that at bottom are self-relations, particles that intrinsically are identical in the strong sense of numerical identity.

Why *formless* particles? What the usual characterization of a fundamental particle as a pointlike entity amounts to is that it lacks internal structure. The notion that a fundamental particle is literally pointlike is warranted neither empirically nor theoretically. In addition it is inconsistent with the incomplete spatial differentiation of the physical world.

The bottom line: Instead of being put together, the world is *manifested*. If I conceive of space as the totality of existing spatial relations, and of matter as the totality of existing particles, I am in a position to affirm that a single transcendent Being brings into existence both matter and space simply by entering into spatial relations with itself. The multitude of particles in existence is then nothing but the multitude of relata that is implied by the existence of spatial relations. Because the spatial relations that obtain between particles are self-relations, this multitude is effective rather than fundamental. Ultimately there is nothing but Being and relations between Being and Being.

This transcendent Being, to which quantum physics has been trying to draw our attention for nearly a century, is the knot that ties together consciousness and matter. But before I come to the second part of my paper, I still need to attend to an apparent logical circle. On the one hand, macroscopic objects are made of microscopic objects. On the other, macroscopic objects are needed to realize the properties of microscopic objects. How can that be?

The key to the resolution of this apparent circle is that the theoretical description of a microscopic object (such as an atom) is a description not in terms of actually measured properties but in terms of correlations between the possible outcomes of measurements that are not actually made. Our understanding of how fundamental particles constitute nucleons, nuclei, atoms, and molecules does not require that fundamental particles possess actually measured properties. Nor does our understanding of the instrumentality of microscopic objects in the manifestation of the macroworld require the attribution of actually measured properties to microscopic objects. The fact that microscopic *observables* need to be measured in order to possess values therefore in no wise prevents microscopic *objects* from playing the role that they do in the manifestation of the macroworld.

Part II

6 Consciousness: "Involution"

So how is consciousness present in that transcendent Being, and how is it present in each fundamental particle? My views on this matter are indebted to the Indian philosopher (and freedom fighter, and mystic) Sri Aurobindo (2005; Heehs 2008). In keeping with a more than millennium-long philosophical tradition (Phillips 1995), Sri Aurobindo posits an Ultimate Reality whose intrinsic nature is (objectively speaking) infinite Quality and (subjectively speaking) infinite Delight. This has the power to manifest its inherent Quality/Delight in finite forms, and the closest description of this manifestation is that of a consciousness creating its own content.

In the native poise of this consciousness, its single self is coextensive with its content and identical with the substance that constitutes the content. This self is wherever its objects are. We may call this "the view from everywhere".

A first self-modification of this *supramental* consciousness leads to a poise in which the self adopts a multitude of standpoints, localizing itself multiply within the content of its consciousness and viewing it in perspective. It is in this secondary poise that the dichotomy between subject and object, or self and substance, becomes a reality.

Probably the most adequate description of the process by which the self assumes a multitude of standpoints is that of a multiple concentration of consciousness. A further self-modification of the original consciousness occurs when this multiple concentration becomes exclusive. We all know the phenomenon of exclusive concentration, when consciousness is focused on a single object or task, while other goings-on are registered subconsciously, if at all. A similar phenomenon transforms individuals who are conscious of their essential mutual identity into individuals who have lost sight of this identity and, as a consequence, have lost access to the supramental "view from everywhere". Their consciousness is mental, which in Sri Aurobindo's terminology means, among other things, that it is concerned with the formation of expressive ideas. Although it receives the quality or qualities it serves to express from a source of which it is no longer aware, it nevertheless commands a wholly effective executive force. This consciousness is closer to the one we are familiar with, but it does not suffer from the compromising consequences of an evolutionary past.

Yet another self-modification of the original consciousness produces individuals who are concerned with execution rather than idea-formation, individuals who receive even the ideas they serve to execute from a subliminal source. And when the multiple exclusive concentration of consciousness is carried to its logical conclusion, it results in individuals who lack even the power of executing ideas. And since this power is responsible for the existence of individual forms, the result is a multitude of formless individuals. We call them elementary particles, and we tend to think of them as the fundamental constituents of matter.

7 Why?

While quantum physics tells us how the probabilities of the possible outcomes of measurements are correlated, it offers no clue to the mechanism or process by which measurement outcomes determine the probabilities of measurement outcomes. What is more, such explanations appear to be ruled out by a growing number of "no-go theorems" (Bell 1964, 1966; Kochen and Specker 1967; Greenberger et al. 1989; Mermin 1985, 1990, 1993; Klyachko et al. 2008). If the force at work in the world is an *infinite* force, this should be no cause for concern, for it would be self-contradictory to explain the working of such a force in terms of physical mechanisms or natural processes. If this force works under self-imposed constraints, what we need to know is why it does so, and why under one particular set of constraints rather than another.

So why would an infinite consciousness render its powers of creation and cognition latent in a multitude of formless particles, and why do the spatial relations between these particles appear to be governed, at least effectively, by general relativity and the theories that make up the Standard Model of particle physics? Here is Sri Aurobindo's answer to the first question:

> a play of self-concealing and self-finding is one of the most strenuous joys that conscious being can give to itself, a play of extreme attractiveness. There is no greater pleasure for man himself than a victory which is in its very principle a conquest over difficulties, a victory in knowledge, a victory in power, a victory in creation over the impossibilities of creation [...] There is an attraction in ignorance itself because it provides us with the joy of discovery, the surprise of new and unforeseen creation [...] If delight of existence be the secret of creation, this too is one delight of existence; it can be regarded as the reason or at least one reason of this apparently paradoxical and contrary Lila (2005, 426–427).

Lila is a term of Indian philosophy that describes the manifested world as the field for a joyful sporting game made possible by self-imposed limitations.

So how is it that our best descriptions of these limitations are Einstein's theory of gravity and the Standard Model? As I have argued in a couple of papers (Mohrhoff 2002, 2009a) and in my book (2011), these theories formulate preconditions for an evolving manifestation of the infinite Quality/Delight that is the very nature of Being. Quantum physics in particular is entailed by the fact that the objects of everyday experience (i) occupy space, (ii) are manifested by means of finite numbers of objects that do not occupy space, and (iii) neither collapse nor explode as soon as they are created. The existence of such objects is an obvious precondition for an evolutionary manifestation of Being, and the fact that such objects are manifested by means of formless particles (which obviously do not occupy space) is a consequence of the *involution* of the ideative and executive powers of Being in formless particles.

8 Consciousness: Evolution

By "evolution" Sri Aurobindo means neither "descent with modification" nor the Darwinist explanation of this historical fact. Essentially, evolution consists in the gradual reversal of the exclusive concentration of consciousness that culminated in the creation of matter. But evolution does not simply retrace the steps that led to the creation of matter, for if it had done so, particles would have acquired forms. What happened instead is that spatial relations between formless particles came to manifest forms. Instead of reversing the transition from formed to formless individuals, evolution uses the outcome of this transition to manifest what has been lost in the transition.

When life appears, what is essentially added to formed individuals is the power to execute ideas, and when mind appears, what is essentially added to living organisms is the power of idea formation. What has yet to evolve is a consciousness that is not exclusively concentrated in the individual, a consciousness aware of the essential mutual identity of all individuals, a consciousness no longer confined to the perspectival outlook of a localized individual but capable of integrating its perspectival outlook with the supramental "view from everywhere".

What about our brains? What necessitated their evolution, and what roles do they play in cognition and volition?

Brains did evolve for a *purpose*: the release of consciousness from its latency in formless particles. Whatever Darwinists may say, this cannot be accomplished without purposeful modifications of the correlations that are encapsulated in our physical theories. As I have argued in my contribution to *The Volitional Brain* (Mohrhoff 1999), such modifications cannot be effected through the loopholes offered by quantum-mechanical indeterminism, without breaching physical laws. Nonphysical influences that preserve the quantum-mechanical probability distributions are causally ineffective, and influences that alter the probability distributions predicted by quantum mechanics breach physical laws. What nevertheless saves the appearances for physicists and Darwinists alike, at least for the time being, is the Houdiniesque nature of this evolving manifestation, for it still limits the scope of purposeful modifications to such an extent that no experiment can reveal statistically significant departures from the physically determined behaviour of matter or from a purely Darwinian mechanism of evolution.

The Houdiniesque nature of this manifestation also explains why something as complex as the human brain had to evolve. Evolution was not meant to be a rapid transformation scene. As yet only very weak nonphysical influences are consistent with the constraints imposed by "this apparently paradoxical and contrary Lila". The complexity of the brain is needed to make such influences physically effective, be it via the brain's trillions of synaptic interconnections or via the intricate system of neural oscillations that it supports. There can be a cumulative non-physical action that contributes to determine the brain's "default" mode of operation, as well as a concerted synchronic action that modifies the "default" mode.

The possibility of nonphysical influences modifying the physical laws is readily understood: an infinite force that can subject itself to the laws of physics can also modify them. But are we the initiators of such influences, and if so, in what sense? Those who believe in a genuinely free will—not compatible with determinism—are likely to attribute it to the intentions or volitions of our waking conscious selves. Most systems of Indian philosophy disagree. They share the fundamental distinction between a deterministic nature (*Prakriti*) and a self (or *Purusha*) the extent of whose freedom varies from one philosophical system to another. Prakriti, which includes not only our bodies but also our minds, evolves deterministically. Purusha, identifying himself with certain physical and certain mental operations of Prakriti, wrongly believes that *he* chooses when in fact *she* (Prakriti) chooses. But the Purusha is also capable of becoming aware of his independent identity and of adopting the attitude of a detached witness, who experiences thoughts, feelings, and actions impersonally and undistorted by any sense of authorship, ownership, or responsibility.

This attitude—a widely acknowledged foundational spiritual experience (Blackmore 1986; Bricklin 1999; Claxton 1999)—is the result of a first reversal of the exclusive concentration by which most of us are confined to their surface waking selves. Adopting it, we can become aware of the actual determinants of our thoughts, our feelings, and our actions, and becoming aware of them, we are once and for all disabused of whatever libertarian illusions we may have harboured. Paradoxically, this disillusionment is the first step towards genuine freedom. For the Purusha then finds that Prakriti functions as she does only by his permission. It becomes possible for him to exert an inner control, which has nothing to do with his erstwhile libertarian imaginations. In the words of Sri Aurobindo:

> If the Purusha in us becomes aware of itself as the Witness and stands back from Nature, that is the first step to the soul's freedom; for it becomes detached, and it is possible then to know Nature and her processes and in all independence [...] to accept or not to accept [...]; we can choose what she shall do or not do in us, or we can stand back altogether from her works and withdraw into the Self's spiritual silence, or we can reject her present formations and rise to a spiritual level of existence and from there re-create our existence (2005, 363).

According to the standard argument against free will, the only alternative to determinism is indeterminism. If determinism is true, we are not free, and if indeterminism is true, freedom is tantamount to randomness. According to the standard argument against this argument, rational decision-making takes place in two stages.

The first stage—the generation of possible courses of action—contains the element of randomness implied by the denial of determinism, while the making of deliberate, non-random decisions takes place in the second stage.

If the deep psychological explorations that have shaped the predominant currents in Indian metaphysics can be trusted, there is no randomness in the making of a decision. Once we have learned to step back from our surface waking selves, we can observe what fills the apparent gaps in the phenomenology of rational decision making. In the words of Sri Aurobindo (2005, 552), "we can observe the springs of our thoughts and feelings, the sources and motives of our action, the operative energies that build up our surface personality". There remains no room for randomness. Which means, in particular, that quantum-mechanical indeterminism has nothing to do with decision making.

Yet Sri Aurobindo (2005, 4) also affirms that "the goal of Nature in her terrestrial evolution" is, *inter alia*, "to establish an infinite freedom in a world which presents itself as a group of mechanical necessities". Arguably, the evolution of infinite freedom begins with the illusion of freedom, which even a paramecium may share with us, and it involves a growing genuine freedom, which only begins to unfold when our now subliminal self—the Purusha—ceases to be subliminal and begins to actively control the operations of Prakriti.

At bottom there is only one way that genuine freedom is possible: to be the sole determinant of this evolving manifestation. In our deepest and truest self we are that. We are the sole determinant of the operations of Prakriti, and this is why we can learn to control and transform them. Needless to say, making this identity fully conscious and wholly effective calls for a series of transformations that extend a good distance into our evolutionary future.

It may be asked: what determines the creative imagination of an all-powerful conscious Self that is also the Substance of its creations? Since the intrinsic nature of this Self and Substance is infinite Delight or Quality, the ultimate purpose of creation can only be to experience this Delight and to express this Quality in finite forms. It deserves to be stressed that such a world-conception has the additional advantage of boldly grounding aesthetic quality and ethical value in the very nature of what is ultimately real.

There also is only one way that knowledge is possible. At bottom all knowledge is knowledge by identity. At their origin, subject and object are one, cognition and will are one, the self for which the world exists is one with the substance by which the world exists. When Being localizes itself multiply within the content of its consciousness, knowledge by identity takes the form of direct knowledge, and when an individual's direct knowledge is limited to a direct knowledge of some of its own physical attributes, as is the case with us, knowledge of external objects takes the form of a knowledge mediated by representations. This brings up the issue of intentionality: why do we perceive external objects rather than our own mediating physical attributes? The answer is contained in this key passage:

> In the surface consciousness knowledge represents itself as a truth seen from outside, thrown on us from the object, or as a response to its touch on the sense, a perceptive reproduction of its objective actuality. Our surface mind is obliged to give to itself this account of its knowledge, because [...] it can catch [...] the surface of outward objects [...] but there is no such ready-made opening between itself and its own inner being. Since it is unable to [...] observe the process of the knowledge coming from within, it has no choice but to accept what it does see, the external object, as the cause of its knowledge [...] In fact, it is a hidden deeper response to the contact, a response coming from within that throws up from there an inner knowledge of the object, the object being itself part of our larger self (Sri Aurobindo 2005, 560–561).

In short, as direct knowledge is supported and made possible by an underlying knowledge by identity, so representational knowledge is supported and made possible by an underlying direct knowledge, which belongs to our subliminal selves.

9 Conclusion

I have been asked to address the question: is quantum physics relevant for the philosophy of mind? My answer is: yes and no.

If the spatial differentiation of the world does not "go all the way down", then neither does its temporal differentiation. In this case quantum states cannot be construed as evolving physical states, and the question of what causes quantum states to collapse does not arise. And if our decisions are fully determined by a combination of physical and non-physical factors, quantum-mechanical indeterminism can have no part in our decisions.

What makes quantum physics relevant to the philosophy of mind is the direction of its explanatory arrow—from unity to multiplicity—and what is implied by it: that material forms are manifested by means of reflexive spatial relations between numerically identical relata. But to bring the relevance of these counterintuitive implications of quantum physics to light, I need to invoke the Vedantic description of Ultimate Reality as something that relates to the world in a threefold manner: as Being (*sat*) it constitutes the world, as Consciousness (*chit*) it contains the world, and as an infinite Quality/Delight (*ānanda*) it expresses and experiences itself in the world. And I need Sri Aurobindo's description of the process by which that which is now evolving came to be involved in an apparent multitude of formless entities.

One last remark. Our theoretical dealings with the world are conditioned by the manner in which we experience the world—by what it is like to be a 21st-century human. We tend to ignore that the manner in which humans experience the world has changed and will change (Gebser 1985; Barfield 1965). Our present mode of experience has enabled us to discover much that is relevant to understanding the past, but it offers little by way of a clue to its future transformations. We tend to think of the evolution of consciousness as a successive emergence of new ways of experiencing a world that, intrinsically, is independent of how it is experienced. But such a world does not exist. There are only different ways in which Being manifests itself to itself. A transformed consciousness implies a transformed world. Our very concepts of space, time, and matter are bound up with, are creations of our present mode of consciousness. It is not matter that has created consciousness; it is consciousness that has created matter, first by carrying its multiple exclusive concentration to the point of being *involved* in a multitude of formless particles, and again by *evolving* to our present mode of experiencing the world, for this has given us the ability to integrate images into three-dimensional objects that appear to exist independently of the experiencing subject. Yet the very logic of this evolving manifestation entails that the next mode of experiencing the world will be one in which the subject rather than the object is the primary reality. Seen by this mode, our theoretical dealings with the world may seem as dated as the mythological explanations of the pre-scientific era seem to us.

References

Barfield, O. 1965, *Saving the Appearances: A Study in Idolatry*, New York: Harcourt Brace & World.
Bell, J.S. 1964, "On the Einstein Podolsky Rosen paradox", *Physics* 1: 195–200.
Bell, J.S. 1966, "On the problem of hidden variables in quantum mechanics", *Reviews of Modern Physics* 38: 447–452.
Bell, J.S. 1990, "Against 'measurement'", *Physics World*, August: 33–40.
Blackmore, S.J. 1986, "Who am I? Changing models of reality in meditation", in G. Claxton (ed.), *Beyond Therapy: The Impact of Eastern Religions on Psychological Theory and Practice*, London: Wisdom.
Bricklin, J. 1999, "A variety of religious experience", *Journal of Consciousness Studies*, 6 (8–9): 77–98; in B. Libet, B. Freeman, and K. Sutherland (eds.) 2000, *The Volitional Brain*, Exeter, UK: Imprint Academic.
Claxton, G. 1999, "Whodunnit? Unpicking the 'Seems' of Free Will", *Journal of Consciousness Studies*, 6 (8–9): 99–113; in B. Libet, B. Freeman, and K. Sutherland (eds.) 2000, *The Volitional Brain*, Exeter, UK: Imprint Academic.
Dirac, P.A.M. 1958, *The Principles of Quantum Mechanics*, Oxford: Clarendon Press.
Feynman, R.P., Leighton, R.B., and Sands, M. 1965, *The Feynman Lectures in Physics*, Vol. 3, Boston, MA: Addison-Wesley.
Gebser, J. 1985, *The Ever-Present Origin*, Athens, OH: Ohio University Press.
Greenberger, D.M., Horne, M., and Zeilinger, A. 1989, "Going beyond Bell's theorem", in M. Kafatos (ed.), *Bell's Theorem, Quantum Theory, and Conceptions of the Universe*, Dordrecht, The Netherlands: Kluwer.
Heehs, P. 2008. *The Lives of Sri Aurobindo*, New York: Columbia University Press.
Klyachko, A.A., Can, M.A., Binicioğlu, S., and Shumovsky, A.S. 2008, "A simple test for hidden variables in the spin-1 system", *Physical Review Letters* 101: 020403.
Kochen, S., and Specker, E. 1967, "The problem of hidden variables in quantum mechanics", *Journal of Mathematics and Mechanics* 17: 59–87.
Mermin, N.D. 1985, "Is the Moon there when nobody looks? Reality and the quantum theory", *Physics Today* 38 (4): 38–47.
Mermin, N.D. 1990, "What's Wrong With These Elements of Reality?", *Physics Today* 43 (6): 9–10.
Mermin, N.D. 1993, "Hidden variables and the two theorems of John Bell", *Reviews of Modern Physics* 65 (3): 803–815.
Mohrhoff, U. 1999, "The Physics of Interactionism", *Journal of Consciousness Studies*, 6 (8–9): 165–184; in B. Libet, B. Freeman, and K. Sutherland (eds.) 2000, *The Volitional Brain*, Exeter, UK: Imprint Academic.
Mohrhoff, U. 2002, "Why the laws of physics are just so", *Foundations of Physics* 32 (8): 1313–1324.
Mohrhoff, U. 2009a, "Quantum mechanics explained", *International Journal of Quantum Information* 7 (1): 435–458.
Mohrhoff, U. 2009b, "Objective probability and quantum fuzziness", Foundations of Physics 39 (2): 137–155.
Mohrhoff, U. 2011, *The World According to Quantum Mechanics: Why the Laws of Physics Make Perfect Sense After All*, Singapore: World Scientific.
Phillips, S. 1995, *Classical Indian Metaphysics*, Chicago/La Salle, IL: Open Court.
Sri Aurobindo 2005, *The Life Divine*, Pondicherry: Sri Aurobindo Ashram Publication Department.

Stuart Hameroff
Consciousness, free will and quantum brain biology – The 'Orch OR' theory

1 Introduction – Consciousness and its place in the universe

We know what it is like to be conscious – to have awareness, phenomenal experience (composed of what philosophers term 'qualia'), a sense of 'self', feelings, sensations, emotions, apparent choice and control of actions, memory, a model of the world and one's body, thought, language, and, e.g. when we close our eyes, or meditate, internally-generated images and geometric patterns. But what consciousness actually *is*, how it comes about and its place in the universe remain unknown.

Science generally portrays consciousness as an emergent property of complex computation among brain neurons. In this view, consciousness first appeared during evolution of biological
 nervous systems. On the other hand, some philosophical, spiritual and quantum physical approaches suggest consciousness depends on a fundamental property intrinsic to the universe, and that consciousness has, in some sense, been in the universe all along. Could both views be true?

The very existence of consciousness seems highly unlikely. Cosmologists tell us that if specific values for the twenty or so fundamental numbers which characterize the universe (precise charge and mass of particles, values for gravitational and other constants, etc.) were just slightly different, life and consciousness—at least as we know them—would be impossible. The universe is seemingly 'fine-tuned' for life and consciousness. Why this may be so is approached by several versions of the 'anthropic principle'. In the 'strong' version (Barrow, and Tipler 1986), the universe is somehow compelled to harbor and enable consciousness, as if consciousness were engaged in its development, organizing the universe. The 'weak anthropic principle' (Carter 1974) suggests that only our particular universe is capable of consciousness, and only this one universe, a privileged version of a multitude of universes, can be observed and wondered about. The question again boils down to whether consciousness is intrinsic to the universe, or an emergent property of brain computation.

The conventional wisdom in neuroscience and philosophy tells us consciousness emerges from brain computation, specifically complex synaptic computation among 'integrate-and-fire' ('Hodgkin-Huxley') brain neurons. The foundation for attempts to understand consciousness is that the brain is a computer. Consciousness is a computation. Some proponents further believe that when the brain's computational wiring diagram—the 'connectome'—is unraveled, mapped and replicated in silicon, brain functions including consciousness will be downloaded and recreated (Kurzweil 2013). Consciousness would become a commodity. Huge resources are aimed at 'mapping the brain'.

But consciousness isn't necessarily computation. Physicist Sir Roger Penrose (1989) points out that while computers surpass humans in many information capacities, they don't really 'understand' anything. And as philosopher David Chalmers' (1996) 'hard problem' illustrates, phenomenal 'qualia' like redness, joy, the taste of mustard and the smell of lilac may involve some added feature, some 'funda-mental' entity or process intrinsic to the fine scale structure of the universe, akin to mass, spin or charge, perhaps embedded with fundamental values which work to anthropically optimize the universe for consciousness.

Unable to account for consciousness through strictly neuronal computational approaches, prominent neuroscientist Christof Koch (2012) has appealed to panpsychism, the notion that material particles are endowed with subjectivity, or experiential 'qualia', intrinsic to the universe as a property of matter. But matter itself, at tiny scales, is continuously 'materializing', i.e. reducing, or collapsing to definite states from multiple quantum possibilities. At the scale at which biomolecules govern neuronal activity, the strange laws of quantum mechanics come into play, and materialism is a mirage. Consciousness seems related to the boundary between quantum and materal worlds.

Physical reality is ruled by two sets of seemingly incompatible laws. In our everyday material ('classical') world, Newton's laws of motion, Maxwell's equations, the gas laws and others accurately predict behavior of particles and energy. However at tiny scales, and the size cutoff, or boundary between the two worlds is variable and unknown, the laws of quantum mechanics rule. Particles can exist in multiple locations or states simultaneously ('quantum superposition'), become spatially separated from one another, but remain connected ('entanglement'), and condense into unitary objects ('quantum coherence').

This strangeness isn't observed in our material world. Attempts to measure quantum superpositions cause them to collapse to definite states. The mystery of why this happens, why there exists some boundary, or edge between quantum and classical worlds is known as the 'measurement problem' in quantum mechanics.

Several interesting solutions to the measurement problem have been put forth. Decoherence is the notion that quantum systems which interact with the classical environment are disrupted by thermal interactions. What about isolated quantum systems?

One proposal from the early days of quantum mechanics is that the very act of conscious observation *causes* quantum possibilities to materialize, or reduce to definite states - consciousness 'collapses the wave function' (e.g. Wigner, von Neumann, Stapp). This view is also known as the 'Copenhagen interpretation' due to the Danish origin of Niels Bohr, one of its early proponents. But this view led to a major dilemma about unobserved, isolated quantum systems, as illustrated by Schrödinger's famous thought experiment in which the fate of an isolated cat is tied to a quantum superposition. According to Copenhagen, the cat is both dead and alive until observed by a conscious human. Absurd it was, but the question persists. Why aren't quantum superpositions seen in our material world?

The 'multiple worlds' hypothesis suggests that with each superposition, the universe separates at a fundamental level, each possibility evolving into its own universe (Everett III 1983). Thus there exists an infinite number of co-existing 'parallel universes'. This view has been linked to the weak anthropic principle, in which we live in the one universe, of a multitude of universes, most conducive to life and consciousness.

These approaches are flawed. But each may each hold part of an answer. The Copenhagen/conscious observer approach has its Schrödinger's cat problem, and places consciousness outside science as the external cause of collapse/reduction. But it does directly link consciousness to quantum state reduction.

'Multiple worlds' is untestable, non-falsifiable, energetically unfavorable, and doesn't deal with consciousness. But it does deal with the nature of superposition. It implies that a particle in two places at the same time is equivalent to separation, bifurcation, in the fine scale structure of the universe—spacetime geometry (irrespective of whether the separated spacetimes evolve to their own universes). Each particle location has its own spacetime geometry.

Another proposed solution to the measurement problem with concepts similar to these two features is Penrose 'objective reduction' (OR) in which quantum superpositions evolve by the Schrödinger equation until reaching an 'objective' threshold for reduction, or collapse. Similar to 'multiple worlds', Penrose OR portrays quantum superpositions as spacetime separations (due to alternate curvatures), but are unstable due to properties inherent in spacetime geometry. Before each spacetime branch evolves its own new universe, the separation reaches OR threshold by the uncertainty principle $E_G = h/t$ (EG is the magnitude of separation, h is the Planck-Dirac constant, and t the time at which OR occurs). At that instant, spacetime geometry reconfigures, quantum possibilities choose particular material states, and, according to Penrose, a moment of conscious experience occurs. Penrose OR turns the Copenhagen/conscious observer approach around. Rather than consciousness *causing* collapse/reduction, consciousness *is* collapse/reduction, a process on the edge between quantum and classical worlds.

Generally, OR can be taken as equivalent to decoherence, the process by which a quantum system is said to be disrupted by its random environment. Superposition/separations E_G arising continuously will entangle other such random superpositions and quickly reach OR threshold by $E_G = h/t$. In such cases, the conscious experience would be primitive qualia without cognitive meaning, described as 'proto-conscious', intrinsic to the universe, accompanying OR events ubiquitously shaping material reality. This approach is similar to the 'Ground of Being' concept in Eastern philosophical terms.

OR 'protoconscious moments' are also similar to Buddhist concepts of discrete conscious moments, and to an approach to consciousness as 'occasions of experience' by philosopher Alfred North Whitehead (1929, 1933) who saw consciousness, and the universe, as a process, as sequences of events. Leibniz had 'quantized' reality, describing fundamental 'monads' as ultimate entities, but Whitehead transformed monads into 'actual occasions' occurring in a "basic field of proto-conscious experience". Whitehead occasions of experience are intrinsic to the universe, spatiotemporal quanta, each endowed, usually, with only low level, "dull, monotonous, and repetitious [...] mentalistic characteristics". Abner Shimony (1993) observed how Whitehead 'occasions' resemble quantum state reductions.

How do we get from simple proto-conscious moments, or occasions, to full, rich meaningful consciousness? In panpsychism, simple particles with simple experience must be somehow organized, or combined into a cognitive, meaningful arrangement - the 'combination problem'. Whitehead considered this problem for his 'occasions', or events, rather than particles, and described how 'highly organized societies of occasions permit primitive mentality to become intense, coherent and fully conscious'.

How can Penrose OR events be so organized, and occur in the context of brain function? The Penrose-Hameroff 'Orch OR' theory suggests OR events are 'orchestrated' into full, rich conscious moments. This paper describes how Orch OR can occur in structures called microtubules inside brain neurons, how it addresses the particular issue of free will, and discusses 'brain tuning', the possibility of addressing mental states and disorders through microtubule quantum vibrations. Consciousness is seen as intrinsic to the universe.

2 Where in the brain does consciousness occur?

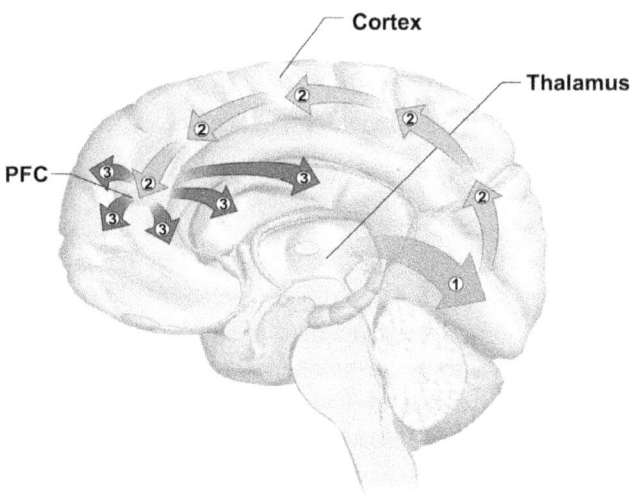

Fig. 1. Three waves in sensory processing. Sensory inputs from spinal cord and cranial nerves to thalamus result in primary projections (1) to primary sensory cortex, e.g. visual area 1 (V1) in occipital cortex in the back of the brain. From these areas, feed-forward projections (2) go to secondary associative and 'executive' areas cortex, e.g. pre-frontal cortex (PFC) from which tertiary projections (3) go to other brain regions whose content then becomes conscious.

The general architecture for conscious sensory processing in the brain is shown in Figure 1. Sensory inputs to thalamus result in (1) projections to primary sensory cortex, e.g. visual area 1 (V1) in occipital cortex in the back of the brain. From primary sensory areas, (2) secondary feed-forward projections go to associative and 'executive' e.g. pre-frontal cortex (PFC). From there, (3) tertiary projections go to other cortical regions whose content then becomes conscious.

The notion that this 'third wave' feedback is conscious, and first and second waves are not conscious, is consistent with philosophical approaches called 'higher order thought' ('HOT'), and neuroscientific cortical feedback models for conscious vision (Lamme, and Roelfsma 2000). Experimental evidence for the association of the 'third wave' with consciousness is provided through studies of anesthesia. Despite the fact that neurotransmitters, receptors and other neurophysiology appears identical among the three waves, all three types of anesthetic molecules (volatile gas anesthetics, propofol and ketamine) selectively inhibit third wave activity while sparing primary and secondary projections (Lee et al, 2013).

There are two clarifications with this anatomical scheme. First, although the brain's medial surface is shown in Figure 3, sensory-based cortical projections may occur more toward outer dorsal surfaces. Second, internally-generated conscious states, e.g. mindwandering, meditation and dreams, possibly mediated through default mode networks, will have different pathways, though their end targets (layer V cortical pyramidal neurons, see below) may be identical.

Third wave activity within cortex seems to also be composed of three waves, successively, and maximally, integrating information. Cortex is arranged in 6 horizontal layers, and sensory inputs from thalamus go (1) to layer 4, and thence (2) from layer 4 to layers 1, 2, 3 and 6. (3) Projections from these layers converge on layer 5 giant pyramidal neurons, the most likely site for consciousness in the brain. Apical dendrites from pyramidal neurons ascend vertically to the cortical surface, and are most directly responsible for measurable electro-encephalography (EEG), e.g. '40 Hz' gamma synchrony, the best neural correlate of consciousness. Axonal firing outputs from layer V pyramidal neurons descend, e.g. to implement behavior, exerting causal efficacy in the world. Third wave integration in cortical layer V pyramidal neurons is the most likely site for consciousness in the brain.

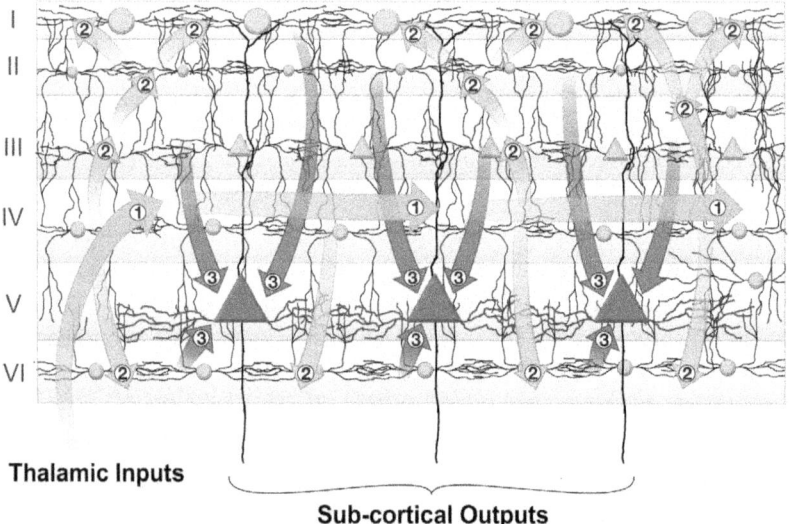

Fig. 2. Three waves of sensory processing in cerebral cortex, a thin mantle on the very top of the brain composed of 6 hierarchical cellular layers. Primary sensory projections from thalamus (1) arrive in layer IV which projects secondary activity (2) to layers I, II, III and VI. These areas then project tertiary (3) activity to giant pyramidal neurons in layer V, where consciousness is most likely to occur. Outputs from layer V pyramidal neurons project sub-cortically, e.g. to manifest 'conscious' behavioral actions. Activity in apical dendrites from pyramidal neurons which ascend to cortical surface are most directly responsible for measurable electro-encephalography (EEG).

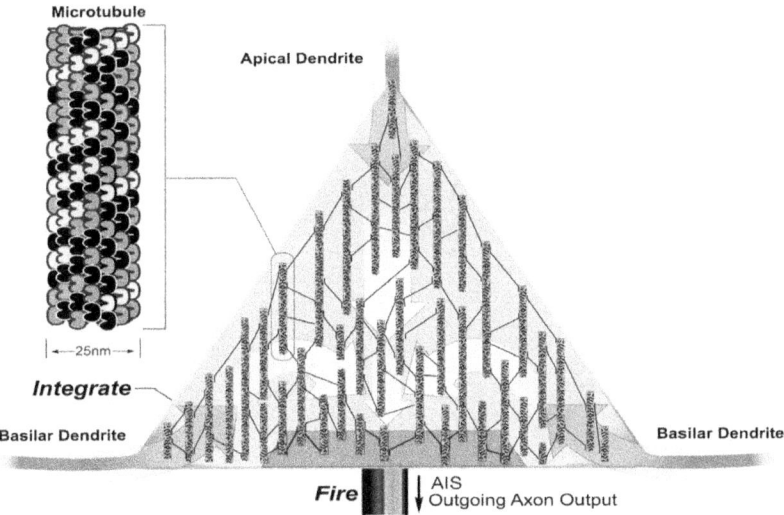

Fig. 3. Layer V pyramidal neuron with internal networks of microtubules connected by microtubule-associated proteins ('MAPs'). Inputs from apical and basilar dendrites are integrated in pyramidal neuronal membranes and cytoskeletal microtubules. On left, a single microtubule is shown comprised of individual tubulin proteins, each in 3 possible states.

'Integrate-and-fire' layer V pyramidal neurons are the final, and maximal, integrator for sensory processing. providing a neurobiological basis for 'Integrated information theory' (Tononi 2012). Their firing outputs control behavior, but neuroscience considers pyramidal neurons (indeed all neurons) according to the Hodgkin-Huxley (HH) standard model. In HH, each neuron is a threshold logic device in which dendrites and cell body (soma) receive and integrate synaptic inputs via excitatory and inhibitory membrane potentials to a threshold at the proximal axon (axon hillock, or axon initiation segment - 'AIS'). When AIS membrane potential reaches a critical threshold, the axon 'fires' to convey signals to the next synapse and layer of neurons.

Integration implies merging and consolidation of multiple disparate information sources. At the level of an individual neuron, integration is approximated as linear summation of synaptic membrane potentials. However integration in branching dendrites and soma requires logic, amplification of distal inputs, branch point effects, and signaling in dendritic spines and local dendritic regions. Nonetheless, according to HH, all such factors are reflected in membrane potentials, and thus the HH neuron is completely algorithmic and deterministic. For a given set of inputs, synaptic strengths and firing threshold, a fixed output in the form of axonal firings, or spikes will occur. Networks of integrate-and-fire neurons regulated by synaptic strengths and firing thresholds can integrate at various anatomical scales, providing highly nonlinear functional processing. But in the end, such processes are algorithmic and deterministic, leaving no apparent room for consciousness or free will.

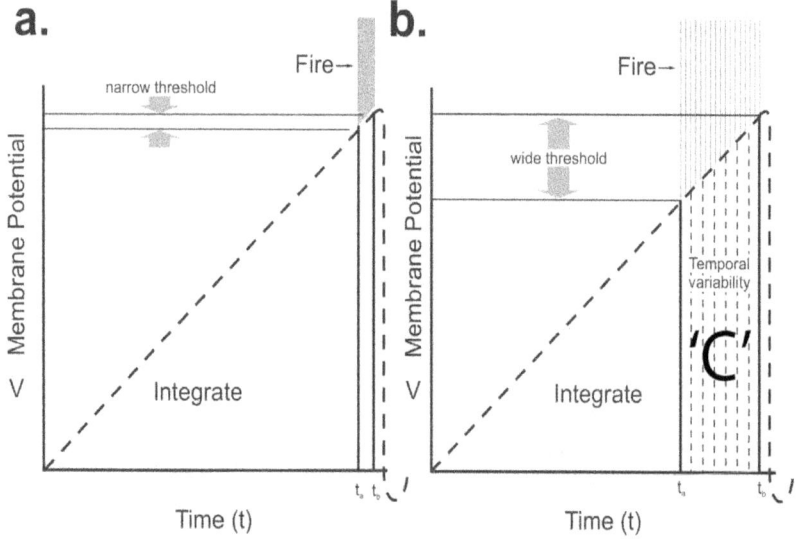

Fig. 4. Integrate-and-fire neuronal behaviors. a. The Hodgkin-Huxley model predicts integration by membrane potential in dendrites and soma reach a specific, narrow threshold potential at the proximal axon (AIS) and fire with very low temporal variability (small t_b-t_a) for given inputs. b. Recordings from cortical neurons in awake animals (Naundorf et al. 2006) show a large variability in effective firing threshold and timing. Some additional factor, perhaps related to consciousness ('C') exerts causal influence on firing and behavior.

However, real neurons differ from idealized HH neurons. For example Naundorf et al. (2006) showed that firing threshold in cortical neurons in brains of awake animals vary spike-to-spike. Some factor other than inputs, synaptic strengths and the integrated membrane potential at the AIS contributes to firing, or not firing. Firings control behavior. This integration 'x-factor' deviation from HH behavior, modulating integration and adjusting firing threshold e.g. in layer V pyramidal neurons, is perfectly positioned for consciousness, causal action and free will, yet is in some way divorced from membrane potentials. What might it be?

3 A finer scale?

Interiors of neurons and other cells are organized and shaped by the cytoskeleton, a scaffolding-like protein network of microtubules, microtubule-associated proteins (MAPs), actin and intermediate filaments.

Microtubules (MTs) are cylindrical polymers 25 nanometers (nm = 10^{-9} meter) in diameter, comprised usually of 13 longitudinal protofilaments, each chains of the protein tubulin. MTs self-assemble from the peanut-shaped tubulin, a ferroelectric dipole arranged within microtubules in two types of hexagonal lattices (A-lattice and B-lattice), each slightly twisted, resulting in differing neighbor relationships among each subunit and its six nearest neighbors. Pathways along contiguous tubulins in the A-lattice form helical pathways which repeat every 3, 5 and 8 rows on any protofilament (the Fibonacci series).

Each tubulin may differ from among its neighbors by genetic variability, post-translational modifications, phosphorylation states, binding of ligands and MAPs, and dipole orientation. MTs are particularly prevalent in neurons (10^9 tubulins/neuron), and uniquely suitable, especially in dendrites and cell bodies, for information processing, encoding and memory. In cell division, MTs dis-assemble, and then re-assemble as mitotic spindles which separate chromosomes, establish daughter cell polarity and then re-assemble for cellular structure and function. However neurons are unlike other cells; once formed, they don't divide, and so neuronal MTs may remain assembled indefinitely, providing a stable potential medium for memory encoding.

MTs in neuronal soma and dendrites are unique in other ways as well. Each tubulin dimer (composed of slightly different alpha and beta monomers) has a dipole, a net positive charge at the beta monomer, and a net negative charge at the alpha monomer. So MTs assembled from parallel arrayed tubulin dipoles also have a net dipole, positive toward its beta monomer end, and negative toward the alpha end. In axons, and in all non-neuronal cells throughout biology, MTs are arrayed radially, like spokes in a wheel, extending continuously from the centrosome near the nucleus, outward toward the cell membrane. These radially arrayed MTs all have the same polarity, the beta plus end outward toward the cell membrane, and alpha negative end inward at the hub, anchored to the centrosome/centriole near the nucleus.

However unlike axons and all other cells, MTs in dendrites and cell bodies/soma are short, interrupted and of mixed polarity, some with their beta plus ends outward, and the rest inward, all interconnected by MAPs into local networks. Dendritic-somatic MTs are also stabilized against depolymerization by special MAP capping proteins, and are thus particularly suitable for memory encoding.

The mechanism by which memory is encoded, stored and processed in the brain remains largely unknown. The standard explanation is through synaptic plasticity, i.e. sensitivities at particular synapses guide activity and create patterns through neuronal networks. However synaptic membrane proteins which determine sensitivity are transient and continuously re-cycled, lasting only hours to days, and yet memories can last lifetimes. Some other factor, or factors, must be involved.

Synaptic proteins are synthesized in neuronal cell bodies/soma, and transported to synapses by 'dynein' and 'kinesin' motor proteins traveling along MTs, which appear to act as passive guides, like railroad tracks. In dendrites and soma where MTs are short, interrupted and of mixed polarity, the motor proteins must jump from MT to MT, and, at dendritic branch points, choose particular pathways to deliver their cargo to the proper synapses. How they do so seems to depend on tau, the microtubule-associated protein (MAP) thought to stabilize MTs, that also seems to serve as traffic signals, instructing motor proteins precisely where to disembark and deliver their cargo. Thus specific placement patterns of tau on MT lattices subserves synaptic function related to memory, and implies specific tau binding locations are encoded in MTs. Tau displacement from MTs results in neurofibrillary tangles, microtubule instability and cognitive dysfunction in Alzheimer's disease.

Thus memory-related synaptic function depends on information encoded in MT lattices, e.g. tau placement. As the origins of memory must reach MTs from the outside world, the question then becomes how synaptic-based inputs may encode information in MTs.

The prevalent synaptic model for memory is long term potentiation (LTP) in which brief, high frequency pre-synaptic stimulation results in long-term post-synaptic potentiation (increased synaptic sensitivity), able to influence neuronal network patterns. At the intra-neuronal level in LTP, synaptic excitation causes influx of calcium ions which convert the hexagonal enzyme calcium-calmodulin to an insect-like calcium-calmodulin kinase II holoenzyme ('CaMKII'). Each of six extended kinase domains on either side of CaMKII are able to phosphorylate (or not phosphorylate) suitable protein substrates, thus providing up to 6 'bits' of information per CaMKII, with hundreds to thousands of CaMKII activation per synaptic excitation. Protein substrates for CaMKII phosphorylations are likely sites for memory encoding, storage and processing. What might they be?

Craddock et al (2012) showed the hexagonal CaMKII kinase array precisely matches hexagonal tubulin lattice spatial geometry in microtubules, and that each kinase domain can reach intra-tubulin amino acids suitable for phosphorylation (Figure 5). CaMKII tubulin phosphorylation may alter dynamical properties, and lead to post-translational modifications resulting in memory 'hardwiring'. Dendritic-somatic microtubules are likely sites for memory encoding.

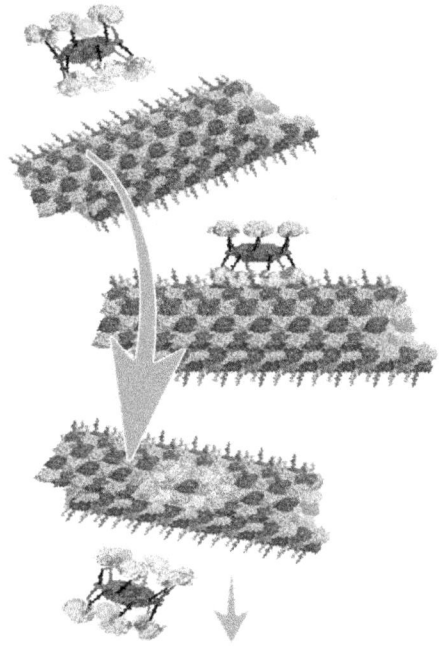

Fig. 5. Memory trace - Calcium-calmodulin kinase II ('CaMKII'), a hexagonal holoenzyme activated by synaptic calcium influx extends 6 leg-like kinase domains above and below an association domain. The 6 kinase domains precisely match hexagonal size and geometry in both A-lattice and B-lattice microtubules.

Due to their lattice structure and organizational roles, MTs have long been suggested to function as information processing devices. Observing and wondering at the intelligent behavior of single cell organisms (lacking synapses), famed biologist Charles Sherrington said: "of nerve there is no trace, but the cytoskeleton might serve" (Adrian 1957). Early descriptions of MTs as computer-like devices (Hameroff, and Watt 1982; Hameroff 1997; Rasmussen et al. 1990) suggested that (1) individual tubulins act as binary bit-like information units (e.g. flexing between two conformations, or dipole states), and that (2) the microtubule lattice acted as a computational matrix or cellular ('molecular') automata. In the latter case, tubulin states interact with hexagonal lattice neighbor tubulin states by dipole couplings, synchronized by biomolecular coherence as proposed by Fröhlich (1968, 1970, 1975; Smith et al. 1984, Rasmussen et al. 1990). Simulations of microtubule automata based on tubulin states show rapid information integration and learning. If the MT memory proposal is correct, information processing relevant to cognition and consciousness would be occurring precisely in the medium in which memory is embedded, a highly efficient and logical proposition.

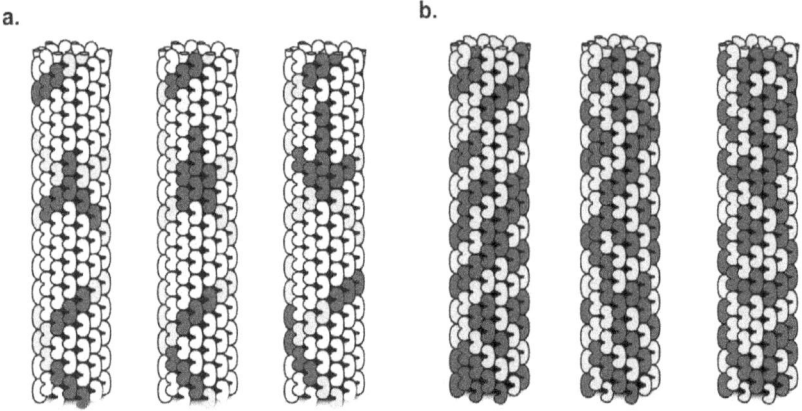

Fig. 6. Three time-steps (e.g. at 10 megahertz) of two types of microtubule automata. (a) Dipole paths or spin currents interact and compute along spiral lattice pathways, for example generating a new vertical spinwave (a 'glider gun' in cellular automata). (b) A general microtubule automata process.

Models of MT information processing developed in the 1980s and 1990s considered fundamental information units to be a bit-like binary state of an individual tubulin, interacting/computing with states of its six surrounding tubulin lattice neighbors. However subsequent models have considered, instead, topological pathways of like tubulin states through two types of MT lattice geometry. In the A lattice with Fibonacci geometry (Figure 7), pathways through adjacent tubulins follow pathways which (1) travel upward/rightward, repeating every 5 tubulins on any single protofilament, and another upward/leftward, repeating every 8 tubulins. Dipole orientations along these pathways may also represent information, interact and compute, perhaps coupled to MT vibrations along these pathways.

In any case, MT-based information processing implies enormous capacity and speed. Based on tubulin binary switching and 10 megahertz processing (see below), MT-based information capacity is roughly 10^9 tubulins per neuron oscillating at e.g. 10 megahertz (10^7 Hz) for 1016 operations per second per neuron.

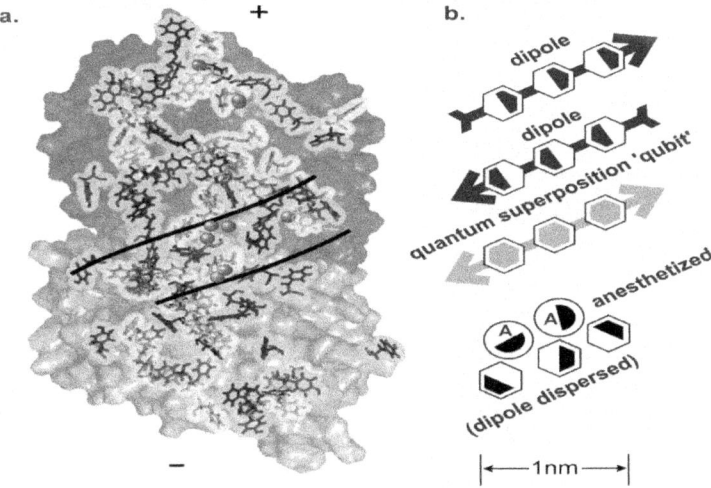

Fig. 7 (a) Molecular modeling of tubulin dimer shows aromatic amino acids tryptophan, phenylalanine and tyrosine in non-polar, hydrophobic regions. Spheres are anesthetic binding sites. Curved lines enclose rings in particular aligned orientation along 5- and 8-start helical channels, containing anesthetic binding sites (with permission from Craddock et al. 2012). (b) Schematic of 5-start helical pathway of aromatic ring dipoles as suggested in Figure 7a. Top 2 pathways show alternate dipole orientations, and 3rd shows quantum superposition of both orientations. Bottom shows how anesthetics disperse dipoles, acting to erase consciousness.

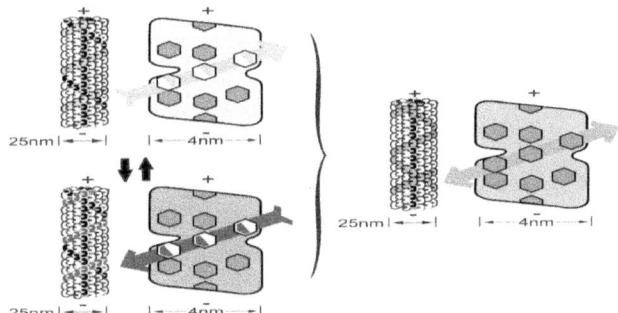

Fig. 8. Dipole qubit in microtubule, with classical and quantum dipole information states for the '5-start' helical pathway in tubulin and microtubules. Left: The '5-start' helix in microtubule A-lattice aligned with dipoles in intra-tubulin aromatic rings (Fig. 7). Top: 'upward' dipole, bottom: 'downward' dipole. Right: Quantum superposition of both upward and downward helical paths coupled to dipole orientations, i.e. 'qubits'. Dipoles may be electric dipoles due to charge separation, or magnetic dipoles, e.g. related to electronic (and/or nuclear) spin. Similar qubit pathways may occur along 8-start pathways, or other pathways.

Dendritic-somatic MTs regulate synapses in several ways. They serve as tracks and guides for motor proteins (dynein and kinesin) which transport synaptic precursors from cell body to distal synapses, encountering, and choosing among several dendritic branch points and many MTs to find the right location. The navigational guidance seems to involve the MAP tau as a 'traffic signal' (specific placement of tau on microtubules being the critical feature). In Alzheimer's disease, tau is hyperphosphorylated and dislodged from destabilized microtubules, forming neurofibrillary tangles associated with memory loss (Matsuyama, and Jarvik, 1989; Craddock et al. 2012a). In Downs syndrome dementia, post-operative cognitive dysfunction (POCD) and other cognitive disorders, MTs are also destabilized and partially disrupted.

Information integration in dendritic-somatic MTs, influenced by encoded memory, may cause deviation from Hodgkin-Huxley neuronal behavior, exerting causal agency. A deeper order, high capacity, finer scale process, e.g. at end-integration in cortical layer V pyramidal neuron dendritic-somatic MTs is a likely site for consciousness. But such a process would still be algorithmic, deterministic, and fail to address phenomenal experience—the 'hard problem'. Something is still missing. Penrose OR provides non-algorithmic (non-computable) processing, causality and addresses the hard problem. In the mid-1990s, Penrose and Hameroff teamed up to apply OR to biology, specifically OR-terminated quantum computations in brain neuronal MTs 'orchestrated' by synaptic inputs, memory and intrinsic MT resonances.

4 Penrose 'Objective Reduction' ('OR') and the 'Orch OR' qubit

Penrose OR is one proposed solution to the 'measurement problem' in quantum mechanics, the problem of why quantum superpositions—particles existing in multiple states or locations at the same time, and described by a quantum wave function—are restricted to microscopic scales, not seen in the 'classical' world we experience. Other suggestions include (1) proposals by Bohr, Wigner, von Neumann, Stapp and others (the 'Copenhagen interpretation', after Niels Bohr's Danish origin) in which conscious observation causes the wave function to collapse (e.g. Schrödinger's cat), but putting consciousness outside science, and (2) decoherence, which suggests that interaction with the random, classical environment disrupts quantum states. (3) 'Multiple worlds' (Everett III 1983) proposes that each possibility evolves its own spacetime geometry, resulting in an infinite number of co-existing universes. And (4), objective reduction (OR) models specify thresholds for quantum state reduction. Among these is Penrose OR.

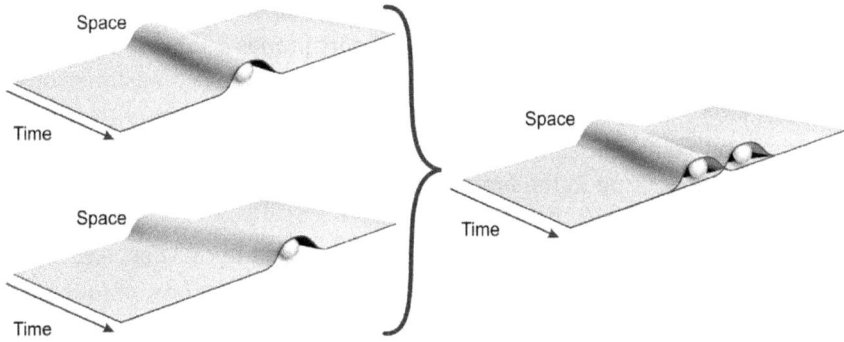

Fig. 9. Spacetime geometry schematized as one spatial and one temporal dimension in which particle location is equivalent to spacetime curvature. Left: Top and bottom show spacetime histories of two alternative particle locations. Right: Quantum superposition of both particle locations as bifurcating space–time depicted as the union ('glued together version') of the two alternative histories (adapted from Penrose 1989, 338).

To approach superposition, Roger Penrose first equated particle states to particular curvatures in spacetime geometry, and superposition to simultaneous, alternate curvatures. Superposition may then be seen as a Planck scale separation, or bubble in the fine scale structure of the universe (Figure 1). In the 'multiple worlds' proposal, each such possible curvature would evolve its own universe. However Penrose suggested spacetime separations were unstable, and would reduce (collapse) due to an objective threshold given by a form of the uncertainty principle $E_G = h/t$. EG is the gravitational self-energy of the superposition, h is the Planck-Dirac constant, and t the time at which OR occurs, accompanied by a conscious moment, and selecting particular states of reality (Figure 2).

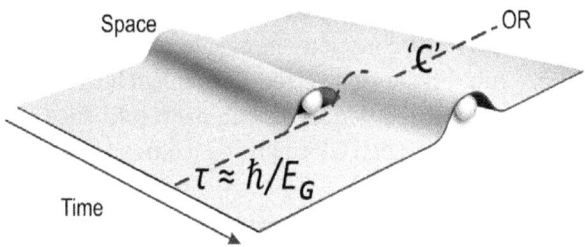

Fig. 10. As superposition curvature EG reaches threshold (by $EG = h/t$), OR occurs. One particle location/curvature is selected and becomes classical. The other ceases to exist.

Generally, such OR events occur in a random environment (identical to decoherence), the accompanying subjective experience lacking cognitive function or meaning. The Penrose-Hameroff 'Orch OR' theory proposes that biology evolved specific mechanisms to isolate and 'orchestrate' OR events ('orchestrated objective reduction' - 'Orch OR'), giving full, rich conscious experience with cognitive meaning and volitional choice. Specific Orch OR mechanisms involve a form of quantum computing in the brain, specifically via structures called microtubules found in all animal and plant cells.

In quantum computers, information is represented not just as, e.g., binary bits of 1 or 0, but also as quantum superposition (quantum bits, or 'qubits') of both 1 and 0. Qubits can entangle, interact and compute with other qubits non-locally, and highly efficiently, according to the Schrödinger equation. (In topological quantum computing, qubits are states of particular pathways through lattices, rather than states of individual subunits comprising those pathways.)

Two basic issues limit implementation of quantum computers. First, interaction with the classical environment disrupts the quantum superposition ('decoherence'), and must be avoided long enough for useful computation to occur. Laboratory quantum computers are hampered by decoherence due to the 'random' thermal environment, and thus constructed at extremely cold temperatures. (Topological qubits are more resistant to decoherence.)

Conceptually, Penrose OR by $E = h/t$ replaces decoherence. Without isolation, in a random environment, OR occurs rapidly, with random outcomes. The combined E_G of the system and its environment quickly reach threshold at h/t, and OR occurs with a non-cognitive, random moment of subjective experience. Presumably, this is occurring continuously, ubiquitously, throughout the universe.

If, however, superposition E_G is isolated from the random environment, 'orchestrated' in a computational register (e.g. a microtubule) by inputs, memory and resonances, and follows the Schrödinger equation to perform cognitive quantum computing, then the orchestrated process will reach OR threshold by $E_G = h/t$, with E_G being composed entirely of orchestrated states. Orch OR then occurs with meaningful cognition and full, rich conscious experience. Sequences of such Orch OR conscious moments provide our familiar 'stream of consciousness'. Tubulin states selected in each Orch OR event (e.g. in layer V pyramidal neuron soma and dendrites) can exert influence on triggering, or not triggering, axonal firing. Orch OR can be the source of 'conscious' deviation from Hodgkin-Huxley behavior.

Orch OR also directly addresses the second issue hampering technological quantum computing, akin to the 'halting problem' in classical computing. If E_G is isolated, premature OR/decoherence is avoided, and the quantum computation proceeds, what then stops it to cause reduction, or collapse to a set of classical values as the solution? In technological quantum computers, the isolated quantum process must, at some point, be 'measured', i.e. the system is observed, causing environmental decoherence, collapse or premature OR. This introduces randomness, and some quantum computers are intended to run the same process repeatedly to average out randomness in measurement/decoherence.

Orch OR offers a direct solution, the quantum computation 'halting' by an objective threshold $E = h/t$. Moreover the specific states (conscious perceptions, actions, tubulin dipoles) selected in each Orch OR event are not randomly chosen, but a product of the quantum computation influenced at the instant of Orch OR by 'non-computable' Platonic factors inherent in spacetime geometry.

Orch OR thus proposes a conscious connection between brain biology and behavior, and the fine scale structure of spacetime geometry through the gravitational self-energy EG of the superposition separation of tubulin in $E_G = h/t$.

According to Orch OR, tubulin states are governed by electronic (and perhaps magnetic) dipoles in non-polar electron clouds, such as aromatic resonance rings of tryptophan, tyrosine and phenylalanine. 32 such rings occur in tubulin, aligned in 'quantum channels', which may align with those in neighboring tubulins, and pathways, in the MT lattice (Figures 7 and 8). These same quantum channels are where anesthetic gas molecules bind by weak, quantum-level London forces to selectively erase consciousness, sparing non-conscious processes. Superposition of tubulin dipole orientations may enable tubulins to act as qubits, and helical pathways through microtubules to act as topological qubits. But electrons have extremely low mass, and E_G for their superposition separation would be very small, requiring extremely long values of t.

However, electron movements of one nanometer shift nearby atomic nuclei by femtometers (Mossbauer recoil and charge attraction), so superposition of electron cloud dipole orientations should result in femtometer superposition separation in tubulin atomic nuclei, sufficient for significant E_G and brief t.

E_G for tubulin superposition separation in Orch OR was calculated in three ways: (1) as separation of 10% of the protein dimer length (~1 nanometer), (2) as separation at the level of the atomic nuclei within each atom of tubulin (femtometer, 10^{-15} meter for carbon), and (3) as separation at the level of nucleons, i.e. protons and neutrons within nuclei (10^{-16} m). The dominant effect was determined to occur at (2) femtometer separation at the level of atomic nuclei. This implies electronic (or magnetic) dipole movements and superpositions in intra-tubulin 'quantum channel' electron cloud resonance rings correspond with femtometer movements and superpositions of nearby atomic nuclei.

Gravitational self-energy E_G of a superpositioned tubulin is then given by $E_G = Gm2/ac$ where G is the gravitational constant, and ac is the superposition separation distance, a carbon nucleus sphere radius equal to 2.5 fermi distances (2.5 femtometers, 2.5 x 10^{-15} meter). If t is assumed to correspond with a neurophysiological event, say 40 Hz gamma synchrony EEG (the best neural correlate of consciousness), then E_G is calculated to be roughly 1010 tubulins. With 109 tubulins per neuron, estimating 0.1% tubulins as quantum coherent (the percent of quantum active molecules in superconductors), requiring 10,000 neurons for t = 25 msec gamma synchrony. But premature OR/decoherence would need to be avoided for 25 msec, a long time in the seemingly 'warm wet and noisy' intra-neuronal environment. Indeed, quantum approaches to brain biology and consciousness have seemed unlikely because of environmental decoherence.

Physicist Max Tegmark (2000) published a critique of Orch OR based on a formula he developed which calculated microtubule decoherence at brain temperature to occur at time tau of 10^{-13} seconds, far from 25 msec. But something was rotten in Tegmark's formula for tau, specifically a term in the denominator for superposition separation. In Orch OR, this is the femtometer diameter of atomic nuclei, however Tegmark described a superpositioned soliton separated from itself by 24 nanometers (3 tubulin lengths) along the microtubule. This gave a decoherence time tau 7 orders of magnitude smaller than it should have been, i.e. from 10^{-13} sec to 10^{-6} sec. Hagan et al. (2001) used Tegmark's same formula, correcting for Orch OR stipulations (superposition separation distance, permittivity etc.) and recalculated microtubule decoherence times to be 10^{-4} sec, suggesting topological resonances would sustain quantum coherence in microtubules for 10^{-1} to 10^{-2} sec. Tegmark's critique, and Hagan et al.'s reply, pitted theory versus theory.

In 2006, experimental research began to show that photosynthesis, the mechanism in plants by which sunlight is converted to chemical energy for food, and without which life could not exist, utilizes quantum coherence at ambient temperatures, i.e. in sunlight. Energy from each photon absorbed in one region of a plant cellular complex is transported as electronic excitations ('excitons') to another region of the complex through a series of 'chromophores', light-absorbing molecules composed of electron resonance clouds. What was surprising—stunning, really—was that the excitons propagated through the chromophores by all possible pathways, essentially a quantum superposition of excitonic pathways. Further work suggested the quantum coherent superposition was aided by coupling to mechanical vibrations in the protein complex. But because the propagation distance was so short, i.e. a few nanometers within the cellular complex, the coherence persisted only for very brief periods of time, e.g. femtoseconds.

What about microtubules? Using nanotechnology, the group of Anirban Bandyopadhyay at the National Institute of Material Science in Tsukuba,

Japan, was able to apply 4 electrodes to a single MT at room temperature. Two electrodes were used to apply very low levels of alternating current (AC) at varying frequencies, and the other two electrodes used to record conductance through the MT. Without AC stimulation, MTs were non-conductive, their resistance being extremely high. However at a number of applied AC frequencies across a wide spectrum (gigahertz, megahertz, and as low as 10 kilohertz), MT resistance dropped, and the MT became significantly conductive (Sahu et al. 2013a, 2013b). More recent studies using nanoprobes inside active neurons also show megahertz and kilohertz coherent vibrations. Particular resonant frequencies may correlate with conductance and vibrations along specific helical pathways through MT lattices.

Quantum resonances as low as 10 kilohertz indicate MT coherence times as long as 10^{-4} seconds, the same MT coherence time calculated by Hagan et al. (2001) using Orch OR stipulations. If t in $E_G = h/t$ is set to $10-4$ secs, E_G of tubulins in 10 million neurons would be required. For 10^{-7} secs, 10 MHz (also proven) E_G of tubulins in 10 billion neurons (or higher involvement per neuron) would be required. Indeed, Orch OR may be seen to occur at different frequency ranges, e.g. akin to different scales in music. But $t = 10^{-4}$ secs is still too brief for physiological effects, such as gamma synchrony EEG at 25 msec.

In Hameroff and Penrose (2014) it was proposed that EEG rhythms are 'beat frequencies' of faster oscillations in microtubules. For example MT megahertz vibrations of slightly different frequencies, or energies, would interfere to give much slower 'beats', e.g. in the 1 to 100 Hz range seen in EEG. Indeed, consciousness may be more like music than computation, sequences of events, at different frequencies, in some cases harmonically related. Quantum vibrations in brain MTs punctuated by Orch OR events are also ripples, or rearrangements in fundamental spacetime geometry. Orch OR connects conscious brain activities to processes in the fine scale structure of the universe.

Each Orch OR conscious moment, e.g. occurring in dendritic-somatic MTs in layer V cortical pyramidal neurons, also selects tubulin states which govern neuronal activities including axonal firing, thus exerting causal action and conscious control of behavior. Can Orch OR account for free will?

5 Free will – Is consciousness too late?

Free will implies conscious agency—that 'we' have conscious causal control and choice of our actions. Indeed, 'we' do seem to have conscious control, but do we really? First, who, or what exactly is 'we', or 'I'? There is no agreed-upon mechanism for consciousness nor conscious agency in neuroscience and philosophy. However Orch OR does offer a mechanism for conscious causal action—tubulin states selected in each Orch OR event may trigger, or not trigger, axonal firings to implement behavior (deviation from Hodgkin-Huxley).

A second issue involves determinism, the notion that all processes in the world, and in our minds, are algorithmic and our choices inevitable and predetermined, with perhaps a dash of randomness. Determinism implies that conscious perceptions and actions follow a complex script written by the laws of nature and history of the universe. Penrose OR avoids determinism by 'non-computable', non-algorithmic influence on selection of particular states at the instant of reduction. According to OR, the quantum wave function of superposition E_G evolves algorithmically according to the Schrödinger equation up until the moment of OR at time t ($E_G = h/t$). At that instant, according to Penrose OR, non-random, 'non-computable Platonic values' embedded in the fine scale structure of spacetime geometry influence choices selected in the OR process. Whether such Platonic values are themselves algorithmic and deterministic are unknown. But to some extent at least, Orch OR dodges determinism.

A third issue with free will pertains to the timing of conscious action, in that consciousness, in some cases, appears to come too late. Neural correlates of conscious perception occur 150 to 500 milliseconds (msec) after impingement on our sense organ, yet we often consciously respond to those perceptions within 100 msec after sensory impingement. For example (Velmans 1991) analysis of sensory inputs and emotional content, phonological and semantic analysis of heard speech, preparation of spoken words and sentences, forming memories, and performing voluntary acts all occur, seemingly consciously, before the stimuli to which the responses were aimed are processed. The conclusion among neuroscientists and philosophers (Dennett, and Kinsbourne 1992; Wegner 2002) has been that we act non-consciously, and have belated, false impressions of conscious causal action. This implies that free will does not exist, that consciousness is epiphenomenal, and that we are, as T.H. Huxley bleakly summarized, "merely helpless spectators".

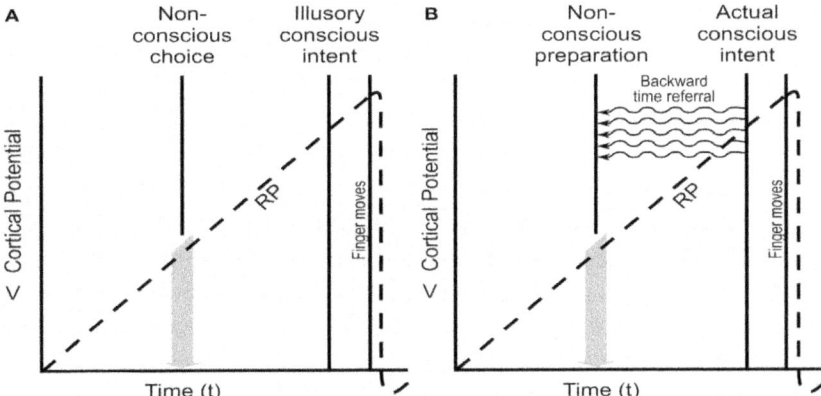

Fig. 11. The "readiness potential (RP)" (Libet et al. 1983). (A) Cortical potentials recorded from a subject instructed to move his/her hand whenever he/she feels ready, and to note when the decision was made (Conscious intent), followed quickly by the finger actually moving. (Time between Conscious intent, and finger moving is fixed.) Readiness potential, RP, preceding Conscious intent is generally interpreted as representing the Non-conscious choice to move the finger, with Conscious intent being illusion. (B) Assuming RP is necessary preparation for conscious finger movement, Actual conscious intent could initiate the earlier RP by (quantum) temporal non-locality and backward time referral, enabling preparation while preserving real time conscious intent and control.

Some evidence appears to support the epiphenomenal contention. Kornhuber and Deecke (1965) recorded electrical activity over pre-motor cortex in subjects who were asked to move their finger randomly, at no prescribed time. Gradually-increasing brain electrical activity preceded finger movement by ~800 msec, termed the *readiness potential* ('RP'). Libet and colleagues (1983) repeated the RP experiment, except they also asked subjects to note precisely when they consciously decided to move their finger. (To do so, and to avoid delays caused by verbal report, Libet et al. used a rapidly moving clock and asked subjects to note when on the clock they consciously decided to move their finger). The 'conscious decision' came ~200ms before actual finger movement, hundreds of milliseconds after onset of the RP. Kornhuber, Deeke, Libet and others concluded the RP represented a non-conscious causal action of the finger movement, that many seemingly conscious actions are initiated non-consciously, and that conscious intent is an illusion. Consciousness apparently comes too late.

But does it? Strangely, apparent backward time effects are observed in brain biology related to consciousness, and theoretically allowable in quantum physics. Could quantum backward time effects rescue conscious free will?

In physics, the 'arrow of time' implies a flow of time in one direction, toward increasing entropy according to the second law of thermodynamics.

However, other laws of physics are generally 'time reversible', working quite well in both directions. In principle, there's no reason for an exclusively unidirectional arrow of time. In the Wheeler-Dewitt equation, which attempts to mathematically reconcile quantum mechanics and general relativity, time plays no role from an external viewpoint. However in the Wheeler-Dewitt equation, conscious observers within the universe do perceive a flow of time and events, the unidirectional flow of time being exclusively related to consciousness (cf. Hameroff 2003). However consciousness may utilize backward time effects.

A principal hallmark of quantum physics is quantum entanglement which implies complementary quantum particles (e.g. electrons in coupled spin-up and spin-down pairs) remain entangled when separated spatially (or temporally). Einstein initially objected, as it would require signaling faster than light, and thus violate special relativity. He famously termed it 'spooky action at a distance', and (with colleagues Podolsky and Rosen—'EPR') described a thought experiment in which an entangled pair of superpositioned electrons (EPR pairs) would be sent in different directions, each remaining in superposition. When one electron was measured at its destination and, say, spin-up was observed, its entangled twin miles away would, according to the prediction, correspondingly reduce instantaneously to spin-down which would be confirmed by measurement. The issue was unresolved at the time of Einstein's death, but since the early 1980s (Aspect et al. 1982, Tittel et al. 1998) this type of experiment has been repeatedly confirmed through wires, fiber optic cables and via microwave beams in the atmosphere. Entanglement is an essential feature of quantum information technologies such as quantum cryptography, quantum teleportation and quantum computers. How does it occur?

Penrose (2004; cf. Bennett, and Wiesner 1992) proposed that measurement and reduction of one twin of the EPR pair sends quantum information backward in time to when the particles were spatially together, then onward to the second twin. According to this scheme, apparent backward time effects are necessary for entanglement, and thus ubiquitous. Aharonov has proposed that quantum state reductions send quantum information both forward and backward in time. In his 1989 book *The emperor's new mind*, Penrose (1989) suggested quantum effects could mediate the brain's backward time referral of subjective information reported by Benjamin Libet.

In the late 1970s and 1980s, Libet and colleagues (in addition to their RP 'move your finger' research, described above) studied the timing of conscious sensory experience in awake, cooperative patients undergoing brain surgery with local anesthesia. This allowed stimulation and recording of somatosensory cortex (e.g. of the hand), periphery (e.g. hand) and direct report of the timing of conscious experience. (To avoid delays due to reporting, subjects observed a fast moving clock, and noted precisely when a sensory perception occurred.) Two types of sensory stimulation were used, one type involving direct stimulation of the skin of the hand, recording the sensory activity from somato-sensory 'hand area' of cortex, and obtaining the subject's report of the precise time of the conscious experience (via the fast-moving clock). Generally, stimulation of the hand resulted in (1) a cortical sensory-evoked potential (EP) at 30 msec after stimulation, and (2) conscious sensory experience also occurring at 30 msec, 30 msec being roughly the time required for neural signals to reach the brain from the hand. In these same subjects, Libet and colleagues also directly stimulated the 'hand area' of somato-sensory cortex. They found no EP, and discovered that 500 msec of continuously-induced cortical activity was required for conscious sensation of the hand to occur at 500 msec. Libet concluded that 500 msec of cortical activity was required to reach threshold for conscious 'neuronal adequacy'. This requirement for several hundreds of msec of direct cortical stimulation to produce conscious experience ('Libet's half second') was subsequently confirmed by Amassian et al. (1991), Ray et al. (1999), Pollen (2004) and others. It was also consistent with the subsequent work of Velmans (1991), Dennett, Kinsbourne and others who maintained that consciousness occurred several hundred msec after sensory impingement, and after seemingly conscious responses. But then, how can conscious experience occur at 30 msec with hand stimulation and the EP?

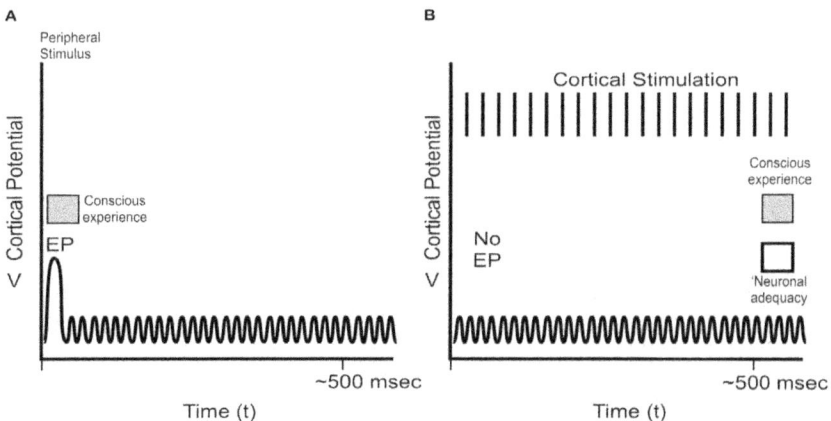

Fig. 12. Cortical potentials in Libet's sensory experiments. (A) Peripheral stimulation, e.g., at the hand, results in near-immediate conscious experience of the stimulation, an evoked potential EP at ~30msec in the "hand area" of somatosensory cortex, and several 100 msec of ongoing cortical electricalactivity. (B) Direct cortical activity of the somatosensory cortical hand area for several 100msec results in no EP, ongoing cortical activity, and conscious sensory experience of the hand, but only after ~500msec. Libet termed the 500msec of cortical activity resulting in conscious experience.

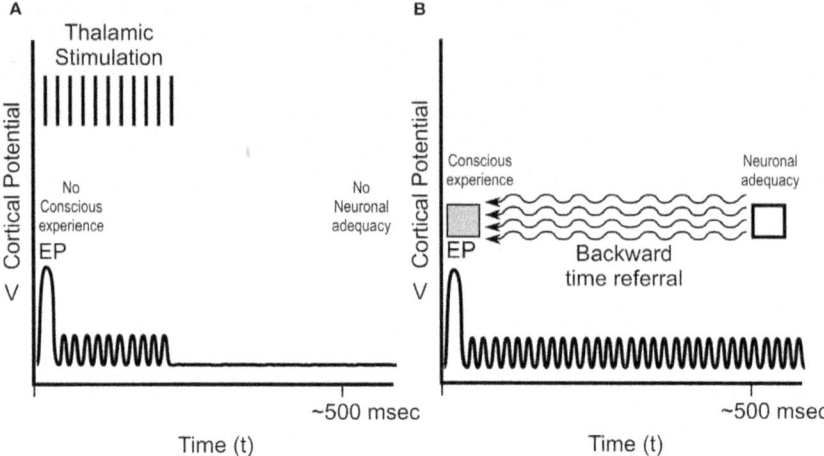

Fig. 13. Libet's sensory experiments, continued.(A) Libet et al. stimulated medial lemniscus of thalamus in the sensory pathway to produce an EP (~30ms) in somatosensory cortex, but only brief post-EP stimulation, resulting in only brief cortical activity. There was no apparent "neuronal adequacy," and no conscious experience. An EP and several100 msec of post-EP cortical activity (neuronal adequacy) were required for conscious experience at the time of EP. (B) To account for his findings, Libet concluded that subjective information was referred backward in time from neuronal adequacy (~500msec) to the EP.

To address this issue, Libet and colleagues did further studies in human subjects stimulating medial lemniscus of thalamus, the primary relay station between, e.g. hand and sensory cortex. Stimulating thalamus, they observed a cortical EP at 30 msec, and ongoing cortical activity for as long as they stimulated thalamus. If the stimulation and ongoing cortical activity persisted for ~500 msec, the subjects reported conscious experience at 30 msec, the time of the EP. If, however, stimulation and cortical activity were stopped after the EP, but prior to ~500 msec, no conscious experience occurred. Somehow, the brain 'knew' whether or not cortical activity would continue for hundreds of msec after the EP for consciousness to occur at the time of the EP. Libet concluded that subjective information was referred 'backward in time' from neuronal adequacy at ~500 msec to the time of the EP at 30 msec. Libet's backward time assertion was disbelieved and ridiculed (e.g. Churchland 1991; Dennett, and Kinsbourne 1992), but never refuted. Indeed, several types of experiments have continued to show backward time effects in the brain.

Electrodermal activity measures skin impedance, usually with a probe wrapped around a finger, as an index of autonomic, sympathetic neuronal activity causing changes in blood flow and sweating, in turn triggered by emotional response in the brain. Researchers Dean Radin and Dick Bierman have published a series of well-controlled studies using electrodermal activity to look for emotional responses to images presented at random times on a computer screen. They found that emotional images elicited responses half a second to two seconds *before* the images appeared. They termed the effect pre-sentiment because the subjects were not consciously aware of the emotional feelings; non-conscious emotional sentiment (i.e. feelings) appeared to be referred backward in time.

In 2011, Daryl Bem published "Feeling the future: Experimental evidence for anomalous retroactive influences on cognition and affect" in the mainstream *Journal of Personality and Social Psychology*. The article reported on 8 studies showing statistically significant backward time effects, most involving non-conscious influence of future emotional effects (e.g. erotic or threatening stimuli) on cognitive choices. Mossbridge et al. (2012) published a meta-analysis of 26 reports published between 1978 and 2010 showing backward time effects, and concluded the results were valid. Moreover they pointed to findings in mainstream neuroscience which show backward time effects but are not reported.

In the famous double slit experiment in quantum physics, quantum entities (e.g. photons, electrons) can behave as either waves, or particles, depending on the method chosen to measure them. John Wheeler described a thought experiment in which the measurement choice (by a conscious human observer) was delayed until after the electron or other quantum entity passed though the slits, presumably as either wave or particle. Wheeler suggested the observer's delayed choice could retroactively influence the behavior of the electrons, e.g. as waves or particles. The experiment was eventually performed and confirmed Wheeler's prediction; conscious choices can affect previous events, as long as the events were not consciously observed in the interim.

In 'delayed choice entanglement swapping', originally a thought experiment proposed by Asher Peres (2000), Ma et al. (2012) went a step further. In entanglement swapping, two pairs of unified/entangled particles are separated, and one from each pair is sent to two measurement devices, each associated with a conscious observer ('Alice' and 'Bob', as is the convention in such quantum experiments). The other entangled particle from each pair is sent to a third observer, Victor. How Victor decides to measure the two particles (as an entangled pair, or as separable particles) determines whether Alice and Bob observe them as entangled (showing quantum correlations) or separable (showing classical correlations). This happens even if Victor decides *after* Alice's and Bob's devices have measured them (but before Alice and Bob consciously view the results). Thus Victor's conscious choice affects behavior of previously measured, but unobserved, events. Entanglement apparently includes not only spooky action at a distance, but spooky action on past events. Anton Zeilinger, senior author on the Ma's et al.

study, said: "Within a naïve classical worldview, quantum mechanics can even mimic an influence of future actions on past events".

Such influences in the brain can allow real-time conscious control of our actions, seen as deviation from Hodgkin-Huxley neuronal behavior (Figure 14). With quantum brain biology, consciousness does not come too late. Free will is possible.

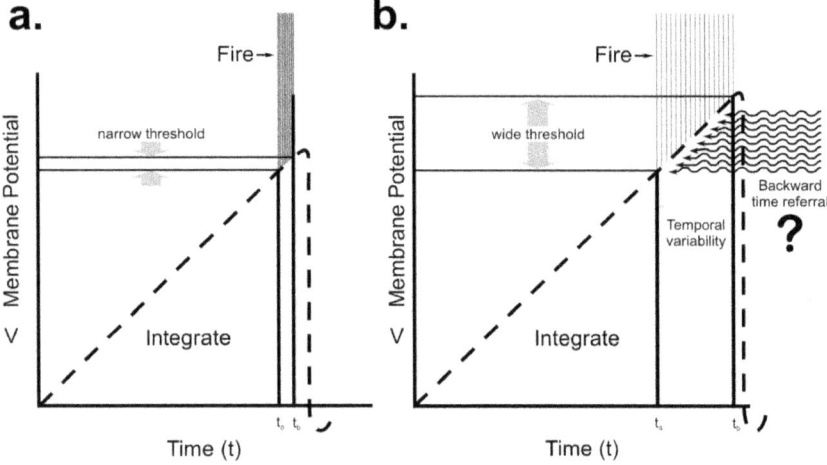

Fig. 14. As in Figure 4, Integrate-and-fire neuronal behaviors. a. The Hodgkin-Huxley model predicts integration, narrow threshold potential and low temporal variability in firing at the proximal axon (AIS) b. Recordings from cortical neurons in awake animals (Naundorf et al. 2006) show a large variability in effective firing threshold and timing. Some additional factor, perhaps related to consciousness ('C') exerts causal influence on firing and behavior, and may include backward time referral.

6 Tuning the brain

Orch OR has been skeptically viewed on the basis of 'decoherence' (i.e. random, 'un-orchestrated' OR). Technological quantum computers require extreme cold, near absolute zero temperature, to avoid thermal vibrations which appear to disrupt delicate quantum effects. Critics said the brain is simply too 'warm, wet and noisy' for functional quantum effects.

Orch OR countered theoretically that coherence akin to Bose-Einstein condensation, and described for biological systems by Herbert Fröhlich, converted thermal energy to coherent vibrations, somewhat like a laser. Quantum spin transfer through aromatic rings was shown to be promoted by heat, not disrupted (Ouyang and Awschalom 2003), and beginning in 2006, evidence began to accrue for warm temperature quantum coherence in photosynthesis proteins (Engel et al, 2007).

Recently, room temperature Bose-Einstein condensation has been demonstrated (Plumhof et al 2014). The brain is not too 'warm' for functional quantum effects.

Orch OR also suggested microtubule quantum coherence originated in isolated, non-polar, 'hydrophobic' regions within tubulin ('quantum channels'), shielded from polar, aqueous interactions. At such quantum sites, anesthetic molecules selectively erase consciousness, acting by quantum London forces. The brain is not too 'wet' for functional quantum effects, at least not in 'dry' quantum channels.

Is the brain too 'noisy'? Seemingly random electrical fluctuations occur continuously throughout the brain, embedded as background in the EEG. The fluctuations emanate mostly from neuronal post-synaptic membrane potentials which don't reach threshold for axonal firing. According to standard neuroscience doctrine, such activity is irrelevant, as firings are all that matter, and dendritic-somatic 'noise' (though it constitutes 'integration') is ignored as irrelevant. Is it really?
The 'noise' is seen at all levels—neuronal, network, and the entire brain. While appearing locally random, brain electrical noise, or 'ongoing activity', is highly synchronized, or correlated (Arieli et al 1996). The fluctuations are precisely the same everywhere. In pyramidal neurons, simultaneous recording of 'noise' in soma and apical dendrite (micron separation) show 'isopotentiality', near-perfect correlation (Yaron-Jakoubovitch et al 2008). Could correlated 'noise' have a function? Perhaps dendritic-somatic 'noise' is essential, or at least related to, cognition and consciousness (Pockett 2000; McFadden 2002). Perhaps brain noise is, as it has been said, "the brain orchestra warming up". Orch OR suggests correlated brain 'noise' originates from deeper level, finer scale quantum vibrations in microtubules inside neurons.

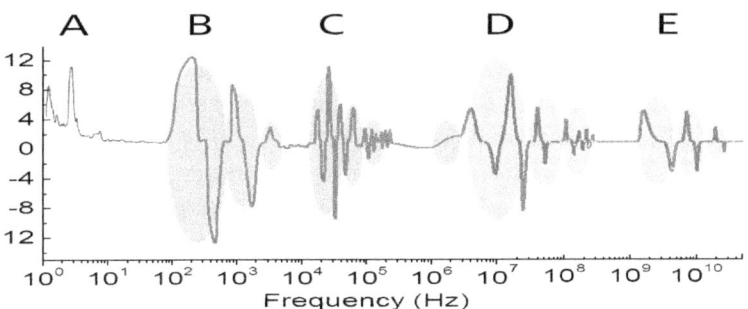

Fig. 15. Five frequency bands of microtubule and brain activity plotted on a log scale. Starting at right, E and D are gigahertz and megahertz resonance frequencies found in individual microtubules (Sahu et al. 2013a, 2013b). B, C and D are kilohertz, tens of kilohertz and megahertz resonance frequencies detected from microtubule bundles inside active neurons (Bandyopadhyay 2014). A is the EEG soectrum. The 5 bands are self–similar and separated evenly by ~3 orders of magnitude, suggesting a harmonic system. EEG (A) may be derived as inverse harmonics, or 'beats' of higher frequency microtubule vibrations.

Brain noise (and measurable EEG) derive from local field potentials due to post-synaptic trans-membrane potentials of roughly 100 millivolt fluctuations, mediated by ion fluxes through membrane protein channels. (Axonal firing potentials, or 'spikes', contribute only in a small way to noise and EEG.) Bandyopadhyay's megahertz and kilohertz electric field fluctuations from microtubule bundles inside neurons of 40 to 50 millivolts are sufficient to influence and regulate membrane potentials. Hameroff and Penrose (2014) suggested interference between microtubules vibrating at slightly different megahertz, or kilohertz, frequencies would give rise to slower 'beat' frequencies, seen as membrane potential fluctuations in EEG or brain noise. EEG is the tip of an iceberg of brain activity.

Thus brain activity relevant to cognition and consciousness may occur at various spatiotemporal scales, moving and combining, like music. Sequences of events at different frequencies, in some cases harmonically related, appear to be anchored by resonances inherent in microtubule lattice geometry. By Penrose OR, Orch OR events are also ripples, or rearrangements in fundamental spacetime geometry. Orch OR connects conscious brain activities to processes in the fine scale structure of the universe.

Within the brain, neuronal and microtubule vibrations span 10 orders of magnitude (Figure 15), and may be directly relevant to mental states. Microtubule vibrations inside brain neurons offer therapeutic opportunities for mood, cognition and neurological disorders.

Modern psychopharmacology aimed at modulating mental states, mood and cognitive function, based on the standard computationalist approach in neuroscience, has as its targets neuronal membrane receptor and channel proteins, and thus may be somewhat misguided. For example the antidepressant Prozac aims to prolong action of the neurotransmitter serotonin at its synaptic receptors (by inhibiting its 'reuptake'). The membrane-mediated effect is immediate, but mood improves only after several weeks, apparently allowing dendritic-somatic microtubules to reorganize (Bianchi et al 2009).

Anti-anxiety benzodiazepine drugs such as Valium, Versed and Xanax are said to act by enhancing binding of GABA (gamma-amino-butyric acid), the brain's primary inhibitory neurotransmitter, to its membrane 'GABA receptor' proteins. But benzodiazepine molecules have several non-polar rings, directly inhibit microtubules in mitosis (Troutt et al. 1995), and are likely to enter neurons and bind in microtubule quantum channels. Similarly, opiate drugs which cause euphoria bind to opiate receptors, but also enter neurons and likely bind in microtubules. While receptor binding occurs, mood-altering drugs may act to tune microtubule vibrations and mellow the music.

Psychedelic drugs are also highly non-polar, contain indole electron resonance rings, and able to enter neurons. Potency of such molecules correlates with their ability to donate electron resonance energy (Kang, and Green 1970; Snyder, and Merril 1965), thus perhaps promoting microtubule quantum vibrations at higher frequencies, and vibrational resonances over many scales.

Anesthetics have opposite effects, selectively erasing consciousness while sparing non-conscious brain functions. They include numerous gas molecules whose potency correlates precisely with solubility in a non-polar, 'olive oil' medium, e.g. as found in lipids, and protein hydrophobic interiors. Franks and Lieb (1984) showed that anesthetics act in non-polar, hydrophobic regions of proteins (not in lipids), presumably membrane proteins, with receptors for acetylcholine, serotonin, glycine and GABA are the most likely candidates. But despite decades of widespread searching, particular membrane receptors or channels mediating anesthetic action have not been found.

In 2006, Rod Eckenhoff's lab at University of Pennsylvania showed that anesthetics such as halothane bind to ~70 proteins in brain neurons, roughly half membrane proteins, and half cytoskeletal proteins including tubulin. Following anesthetic exposure, genetic expression of tubulin, but not of any membrane proteins, was altered. Genomic and proteomic evidence point to microtubules as the site of anesthetic action. Emerson et al. (2013) used fluorescent anthracene as an anesthetic in tadpoles, and showed cessation of tadpole behavior occurs specifically via anthracene anesthetic binding in tadpole brain microtubules. Despite prevailing assumptions, actual evidence suggests anesthetics act on microtubules, not membrane receptors and channels to erase consciousness.

Thus modern psychopharmacology may be aiming at the wrong targets, thus explaining why it isn't more successful in treating mental state disorders. A primary reason membrane receptors and channels are studied is that their effects (drug binding, conformational change, channel opening) are measurable. Until recently, there was no assay for microtubule function other than polymerization states of assembly/disassembly. Now however resonance vibrations, e.g. in megahertz, have been discovered, and drugs may be assayed for their effects on microtubule resonance spectra inside neurons. New vistas await in the study of drug effects on microtubule vibrations.

Another avenue to treating mental states and cognitive dysfunction comes through noninvasive brain stimulation techniques. Among these are transcranial magnetic stimulation ('TMS'), transcranial electrical direct current stimulation ('TDcS') and transcranial ultrasound stimulation ('TUS'), all of which have shown promise and interesting effects. Among these, only TUS can be narrowly focused to target specific, deeper brain regions (Legon et al. 2014).

Ultrasound consists of mechanical vibrations above human hearing threshold (~20,000 Hz), and is usually used in the low megahertz (106 to 107 Hz) for medical imaging, passing through the body and echoing back off surfaces.

TUS consists of low intensity, sub-thermal levels of ultrasound administered at the scalp which safely penetrates skull and reaches the brain sufficiently to be echoed back to provide an image of the brain surface and sulci. As microtubules have megahertz vibrational resonances, TUS with proper settings might be expected to enhance microtubule resonance, and thereby affect microtubule functions related to cognition and mental states. Indeed, focused TUS enhances sensory discrimination in human volunteers (Legon et al, 2014), and unfocused TUS improved mood in chronic pain patients (Hameroff et al. 2013).

At the cellular level in embryonic neurons, ultrasound promotes growth of neurites leading to formation of axons, dendrites and synapses (Raman). At the level of tubulin, ultrasound promotes microtubule assembly. As traumatic brain injury involves disrupted microtubules, synapses and circuits, and as Alzheimer's disease and post-operative cognitive dysfunction (cognitive decline after anesthesia in elderly), TUS may be useful for all these disorders.

Traumatic memory is an important factor in psychotherapy (e.g. post-traumatic stress disorder). Some suggest eliciting a traumatic memory and then over-writing it at that time with a positive memory (Lane et al. 2013). Since synaptic membrane proteins are too short-lived to store and encode memory, and microtubules appear likely to do so, psychotherapy combined with TUS aimed at microtubule vibrations may be optimal.

To erase or over-write traumatic memory, to change the music and re-tune the tubules, combinations of pharmacology, psychotherapy and TUS (e.g. aimed at microtubule vibrations in amygdala, hippocampus and pre-frontal cortex) may be optimal. As the Beatles sang, "Take a sad song and make it better".

7 Conclusion

The mainstream materialist approach to brain function in neuroscience and philosophy suggests that consciousness and cognition emerge as higher order network effects from complex computation among relatively simple neurons. The fine grain of conscious and cognitive information is conveyed at the neuronal level by axonal firings and synaptic transmissions mediated entirely by membrane proteins.

This approach has failed. Neuronal computational networks fail to account for (1) cognitive phenomenal aspects of single cell organisms like *paramecium* which swim, learn, find food and mates and have sex, all without synaptic connections, using their cytoskeletal microtubules for sensory processing and motor actions, (2) phenomenal subjective aspects of consciousness (the 'hard problem'), (3) free will as real-time conscious control, neuronal activity correlating with perception occurring after seemingly conscious response, relegating consciousness to epiphenomenal 'helpless spectator', (4) memory, as membrane proteins determining synaptic sensitivity are transient, and yet memories can last lifetimes, and (5) molecular mechanisms for drugs affecting consciousness including anesthetics, which selectively erase consciousness, but despite popular belief, do not act on membranes, and (6) scientific plausibility for non-locality, e.g. so-called telepathy, pre-cognition, near death/out-of-body experiences, and afterlife. These are generally stated to be impossible because they cannot be scientifically explained by the mainstream view based on neuronal-based computation. But the mainstream materialist view can't really explain anything about consciousness. And materialism itself is illusory, as particles repeatedly coalesce from quantum possibilities.

In spiritual and idealist approaches, consciousness has in some sense always existed in the universe, being in some way intrinsic to its very makeup. Some such views place consciousness as primary, omnipresent, with matter and the world manifesting within an all-pervading consciousness. But such approaches themselves are as yet untestable and unfalsifiable, essentially putting consciousness outside science.

Orch OR is based on deeper level quantum vibrations in microtubules inside neurons, quantum vibrations which occur in the fine scale structure of spacetime geometry. Orch OR provides a bridge between the two approaches, and an opportunity to treat mental disorders by tuning microtubule quantum vibrations.

Acknowledgments

Thanks to Roma Krebs for artwork, Kamber Geary for manuscript preparation and Drs. Antonella Corradini and Mattia Pozzi for their patience and editorial skill.

References

Adrian, Ed. 1957, "Sir Charles Scott Sherrington O.M., 1957-1952", *Notes Rec R. Soc. Lond.* 12: 2011–2015.

Amassian, A. 1991, "Matching Focal and Non-Focal Magnetic Coil Stimulation to Properties of Human Nervous System: Mapping Motor Unit Fields in Motor Cortex Contrasted with Altering Sequential Digit Movements by Premotor-SMA Stimulation", *Electroencephalography and Clinical Neurophysiology*, Supplement 43: 3–28.

Arieli, A., Sterkin A., Grinvald, A., Aersten A. 1996, "Dynamics of Ongoing Activity: Explanation of the Large Variability in Evoked Cortical Responses", *Science* 273 (5283), 1868–1871.

Aspect, A., Grangier, P., Roger, G. 1982, "Experimental Realization of Einstein-Podolsky-Rosen-Bohm Gedanken Experiment: A New Violation of Bell's Inequalities", *Physical Reviews Letters* 48: 91–94.

Bandyopadhyay, A. 2014, "Opening 'Pandora's Box': Direct Measurement of Microtubule Bundle Resonance in a Live Neuronal Axon Suggests Scale-invariant Brain Dynamics Extends Inside Neurons", *The Towards a Science of Consciousness Conference*, Tucson, Arizona April, 21-26, forthcoming.

Barrow, J.D., Tipler, F.J. 1986, *The Anthropic Cosmological Principle*, Oxford: Oxford University Press.

Bem, D.J. 2012, "Feeling the Future: Experimental Evidence for Anomalous Retroactive Influences on Cognition and Affect", *Journal of Personality and Social Psychology* 100, 407–425.

Bennett, C.H., Wiesner, S.J. 1992, "Communication via 1- and 2-particle Operators on Einstein-Podolsky-Rosen States", *Physical Reviews Letters* 69: 2881–2884.

Bianchi, M., Shah, A.J., Fone, K.C., Atkins, A.R., Dawson, L.A., Heidbreder, C.A., et al. 2009, "Fluoxetine Administration Modulates the Cytoskeletal Micro-tubular System in the Rat Hippocampus", *Synapse* 63:359–64.

Carter, B. 1974, "Large Number Coincidences and the Anthropic Principle in Cosmology," *IAU Symposium 63: Confrontation of Cosmological Theories with Observational Data*, Dordrecht: Reidel, 291–298.

Chalmers, D.J. 1996, *The Conscious Mind In Search of a Fundamental Theory*, New York: Oxford University Press.

Churchland, P.S. 1981, "On the Alleged Backwards Referral of Experiences and its Relevance to the Mind-body Problem," *Philosophy of Science* 48: 165–181.

Craddock, T., St George, M., Freedman, H., Barakat, K., Damaraju, S., Hameroff, S, et al. 2012, "Computational Predictions of Volatile Anesthetic Interactions with the Microtubule Cytoskeleton: Implications for Side Effects of General Anesthesia", PLoS ONE 7 (6): e37251.

Craddock, T., Tuszynski, J., Chopra, D., Casey, N., Goldstein, L., Hameroff, S., et al. 2012, "The Zinc Dyshomeostasis Hypothesis of Alzheimer's Disease", PLoS ONE 7 (3): e33552.http://dx.doi.org/10.1371/journal.pone.0033552.

Craddock, T., Tuszynski, J., Hameroff, S. 2012, "Cytoskeletal Signaling: Is Memory Encoded in Microtubule Lattices by CaMKII Phosphorylation?", PLoS Comput Biol 8 (3): e1002421.

Crick, F.C., Koch, C. 2001, "A Framework for Consciousness," *Nature Neuroscience* 6: 119–126.

Dennett, D.C. 1991, *Consciousness Explained*, Boston: Little, Brown.

Dennett, D.C., Kinsbourne, M. 1992, "Time and the Observer: The Where and When of Consciousness," *Behavioral and Brain Sciences* 15:183-247.

Dustin, P. 1985, *Microtubules*, 2nd ed., New York: Springer.
Einstein, A., Podolsky, B., and Rosen, N. 1935, "Can Quantum Mechanical Descriptions of Physical Reality be Complete?", *Physical Review* 47: 777–780.
Emerson, D., Weiser, B., Psonis, J., Liao, Z., Taratula, O., Fiamengo, A., et al. 2013, "Direct Modulation of Microtubule Stability Contributes to Anthracene General Anesthesia", *J Am Chem Soc* 135 (14): 5398.
Engel, G.S., Calhoun, T.R., Read, E.L., Ahn, T.K., Mancal, T., Cheng, Y.C., et al. 2007, "Evidence for Wave-like Energy Transfer Through Quantum Coherence in Photosynthetic Systems", *Nature* 446: 782–786.
Everett III, H. 1983, "Relative State Formulation of Quantum Mechanics", in J.A. Wheeler, and W.H. Zurek (eds.), *Quantum Theory and Measurement*, Princeton: Princeton University Press. Originally in *Reviews of Modern Physics* 29 (1957): 454–462.
Franks, N.P., and Lieb, W.R. 1982, "Molecular Mechanisms of General Anesthesia", *Nature* 316: 349–351.
Fröhlich, H. 1968, "Longrange Coherence and Energy Storage in Biological Systems", *International Journal of Quantum Chemistry* 2: 641–649.
Fröhlich, H. 1970, "Long-range Coherence and the Actions of Enzymes", *Nature* 228: 1093.
Fröhlich, H. 1975, "The Extraordinary Dielectric Properties of Biological Materials and the Action of Enzymes", *Proceedings of the National Academy of Sciences* 72: 4211–4215.
Hagan, S., Hameroff, S., Tuszynski, J. 2002, "Quantum Computation in Brain Microtubules? Decoherence and Biological Feasibility", *Physical Reviews E* 65: 061901.
Hameroff, S.R., and Watt, R.C. 1982, "Information Processing in Microtubules", *J Theor Biol* 98: 549–561.
Hameroff, S.R., and Penrose, R. 1996, "Conscious Events as Orchestrated Space-time Selections," *Journal of Consciousness Studies* 3 (1): 36–53. http://www.u.arizona.edu/~hameroff/penrose2.
Hameroff, S.R., and Penrose, R. 1996, "Orchestrated Reduction of Quantum Coherence in Brain Microtubules: A Model for Consciousness", in S.R. Hameroff, A.W. Kaszniak, and A.C. Scott (eds.), *Toward a Science of Consciousness: The First Tucson Discussions and Debates*, Cambridge: MIT Press, 507–540. Also published 1996 in *Mathematics and Computers in Simulation* 40:453-480. http://www.consciousness.arizona.edu/hameroff/or.html.
Hameroff, S. 2003, "Time, Consciousness and Quantum Events in Fundamental Space-time Geometry", in R. Buccheri, and Saniga M., *The Nature of Time: Physics, Geometry and Perception*, Proceedings of a NATO Advanced Research Workshop.
Hameroff, S., Trakas, M., Duffield, C., Annabi, E., Gerace, M.B., Boyle, P. et al. 2013, "Transcranial Ultrasound (TUS) Effects on Mental States: A Pilot Study", *Brain Stimul* 3 (6): 409–15.
Hameroff, S.R., and Penrose, R. 2014, "Consciousness in the Universe – Review of the Orch OR Theory", *Phys Life Rev*. http://www.sciencedirect.com/science/article/pii/S1571064513001188.
Hayes, D., Griffith, G.B., Engel, G.S. 2013, "Engineering Coherence Among Excited States in Synthetic Heterodimer Systems", *Science* 340 (6139): 1431–1434.
Hildner, R., Brinks, D., Nieder, J.B., Cogdell, R.J., van Hulst, N.F. 2013, "Quantum Coherent Energy Transfer Over Varying Pathways in Single-light Harvesting Complexes", *Science* 340 (639): 1448–1451.

Hodgkin, A., Huxley, A. 1952, "A Quantitative Description of Membrane Current and its Application to Conduction and Excitation in Nerve", *J Physiol* 117: 500–544; http://dx.doi.org/10.1371/journal.pcbi.1002421.

Huxley, T.H. 1893, 1986, "Method and Results: Essays", London: Macmillan.

Kang, S., and Green J.P. 1970, "Steric and Electronic Relationship among Some Hallucinogenic Compounds", *Proc Natl Acad Sci USA* 67 (1): 62–67.

Koch, C., and Crick, F.C.R. 2001, "The Zombie Within", *Nature* 411: 893.

Koch, C. 2004, *The Quest for Consciousness: A Neurobiological Approach*, Englewood: Roberts and Co.

Koch, C. 2012, *Confessions of a Romantic Reductionist*, Cambridge MA: MIT Press.

Kornhuber, H.H., and Deecke, L. 1965, „Hirnpotential Andrugen Bei Willkurbewegungen und Passiven Bewungungen des Menschen: Bereitschaftspotential und Reafferente Potentiale Pflugers", *Archiv* 284: 1–17.

Kurzweil, R. 2013, *How to Create a Mind. The Secret of Human Thought Revealed*, New York: Viking Press.

Lamme, V.A.F., and Roelfsema P.R. 2000, "The Distinct Modes of Vision Offered by Feed-forward and Recurrent Processing," *Trends Neurosci* 23 (11): 571–579.

Lane, R.D., Ryan, L., Nadel, L., Greenberg, L. 2013, "Memory Reconsolidation, Emotional Arousal and the Process of Change in Psychotherapy", *New Insights from Brain Science Behavioral and Brain Science*", Forthcoming.

Lee, U., SeungWoo, M.D., Noh, G., Baek S., Choi, B., Mashour, G. 2013, "Anesthesiology", *Perioperative Medicine* 118 (6): 1264–1275.

Legon, W., Sato, T.F., Opitz, A., Mueller, J., Barbour, A., Williams, A., and Tyler, W.J. 2014, "Nature", *Neuroscience* 17: 322–329.

Libet, B., Alberts, W.W., Wright, W., Delattre, L., Levin, G., and Feinstein, B. 1964, "Production of Threshold Levels of Conscious Sensation by Electrical Stimulation of Human Somatosensory Cortex", *Journal of Neurophysiology* 27: 546–578.

Libet, B., Alberts, W.W., Wright, E.W., and Feinstein, B. 1967, "Response of Human Somatosensory Cortex to Stimuli below Threshold for Conscious Sensation", *Science* 158: 1597–1600.

Libet, B., Wright, E.W., Jr., Feinstein, B., Pearl, D.K. 1979, "Subjective Referral of the Timing for a Conscious Sensory Experience", *Brain* 102: 193–224.

Libet, B., Gleason, C.A., Wright, E.W., and Pearl, D.K. 1983, "Time of Conscious Intention to Act in Relation to Onset of Cerebral Activity (Readiness Potential): The Unconscious Initiation of a Freely Voluntary Act", *Brain* 106: 623–642.

Ma, Q., Tipping R.H., and Lavrentieva N.N. 2012, "Theoretical Studies of N2-Broadened Half-Widths of H2O Lines Involving High States", *Molec. Phys.* 110, 307–331.

Matsuyama, S., Jarvik, L. 1989, "Hypothesis: Microtubules, A Key to Alzheimer Disease", *Proc Natl Acad Sci USA* 86 (20): 8152–8156.

McFadden, J. 2002, "The Conscious Electromagnetic Field Theory: The Hard Problem Made Easy", *J. Conscious. Stud.* 9 (8): 45–60.

Mossbridge, J., Tressold, P., Utts, J. 2012, "Predictive Physiological Anticipation Preceding Seemingly Unpredictable Stimuli: A Meta-analysis", *Front Psychol* 17; doi:10.3389/fpsyg.2012.00390.

Naundorf, B., Wolf, F., and Volgushev, M. 2006, "Unique Features of Action Potential Initiation in Cortical Neurons", *Nature* 440: 1060–1063.

Ouyang, M., Awschalom, D.D. 2003, "Coherent Spin Transfer Between Molecularly Bridged Quantum Dot", *Science* 301: 1074–1078.

Penrose, R. 1989, *The Emperor's New Mind*, 1st ed., Oxford: Oxford University Press.
Penrose, R. 1994, *Shadows of the Mind: A Search for the Missing Science of Consciousness*, Oxford: Oxford University Press.
Penrose, R., and Hameroff, S.R. 1995, "What Gaps? Reply to Grush and Churchland", *J Conscious Stud* 2: 98–112.
Penrose, R. 1996, "On Gravity's Role in Quantum State Reduction", *General Relativity and Gravitation* 28 (5): 581–600.
Penrose, R. 2004, *The Road to Reality: A Complete Guide to the Laws of the Universe*, London, Jonathan Cape.
Peres, A. 2000, "Delayed Choice for Entanglement Swapping", *J. Mod. Opt.* 47, 531; arXiv: quant-ph/9904042.
Plumhof, J.D., Stoferle, T., Mai, L., Scherf, U., and Mahrt, R.F. 2014, "Room-temperature Bose-Einstein Condensation of Cavity Exciton-polaritons in a Polymer", *Nature Materials* 13: 247–252; doi: 10.1038/nmat3825.
Pockett, S. 2000, *The Nature of Consciousness: A Hypothesis*, New York: iUniverse.
Pollen, D.A. 2004, "Brain Stimulation and Conscious Experience", *Consciousness and Cognition* 13 (3), 626–645.
Rasmussen, S., Karampurwala, H., Vaidyanath, R., Jensen, K., and Hameroff, S. 1990, "Computational Connectionism within Neurons: A Model of Cytoskeletal Automata Sub-serving Neural Networks", *Physica D* 42: 428–449.
Ray P.G., Meador, K.J, Smith, J.R., Wheless J.W., Sittenfeld M., and Clifton G.L. 1999, "Physiology of perception. Cortical stimulation and recording in humans", *Neurology* 52 (5): 1044.
Sahu, S., Ghosh, S., Ghosh, B., Aswani, K., Hirata, K., Fujita, D., et al. 2013, "Atomic Water Channel Controlling Remarkable Properties of a Single Brain Microtubule: Correlating Single Protein to its Supramolecular Assembly", *Biosens Bioelectron* 47:141–148.
Sahu, S., Ghosh, S., Hirata, K., Fujita, D., and Bandyopadhyay, A. 2013, "Multi-level Memory-switching Properties of a Single Brain Microtubule", *Applied Physics Letter* 102: 123701.
Shimony, A. 1993a, "Events and processes in the quantum world", in *Search for a Naturalistic World View*, Vol. 2, *Natural Science and Metaphysics*, Cambridge, UK: Cambridge University Press, 140–162.
Shimony, A. 1993b, "The Transient *Now*", in *Search for a Naturalistic World View*, Vol. 2, *Natural Science and Metaphysics*, Cambridge, UK: Cambridge University Press, 271–287.
Smith, S.A., Watt, R.C., and Hameroff S.R. 1984, "Cellular automata in cytoskeletal lattices", *Physica D: Nonlinear Phenomena* 10 (1–2): 168–174. Stapp, H.P. 1993, *Mind, Matter and Quantum Mechanics*. Berlin: Springer.
Snyder, S.H, and Merril, C.R. 1965, "A Relationship between the Hallucinogenic Activity of Drugs and their Electronic Configuration", *Proc Natl Acad Sci USA*, 54 (1): 258–266.
Tegmark, M. 2000, "The Importance of Quantum Decoherence in Brain Processes", Phys. Erd. E 6, 4194–4196.
Tittel, W., Brendel J., Zbinden H., Gisin N. 1998, "Violation of Bell Inequalities by Photons More Than 10 Km Apart", *Phys. Rev. Lett* 81, 3563–3566.
Tononi, G. 2012, *PHI: A Voyage from the Brain to the Soul*, Pantheon Books: New York.
Troutt, L.L., Spurck, T.P., and Pickett-Heaps, J. 1995, "The Effects of Diazepam on Mitosis and the Microtubule Cytoskeleton II. Observations on Newt Epithelial and PtK1 Cells", *Protoplasma* 189 (1,2): 101–112.

Velmans, M. 1991, "Consciousness from a First-Person Perspective", Behavioral and Brain Sciences 14 (4), 702–726.
Wegner, D. 2002, The Illusion of Conscious Will, Cambridge MA: Mit Press.
Whitehead, A.N. 1929, *Process and Reality*, New York: Macmillan.
Whitehead, A.N. 1993, *Adventure of Ideas*, London: Macmillan.
Yaron-Jakoubovitch A., Jacobson G.A., Koch, C., Segev, I., and Yarom, Y. 2008, "A Paradoxical Isopotentiality: A Spatially Uniform Noise Spectrum in Neocortical Pyramidal Cells", *Front Cell Neurosci* 13 (2): 3; doi: 10.3389/neuro.03.003.2008 (eCollection 2008).

Massimo Pauri
Physics, Free Will, and Temporality in the Open World

> Neither the subtlest philosophy nor common human reason can jeopardize freedom by means of sophisms. Reason must therefore assume that no real contradiction exists between freedom and the natural necessity of human actions; because reason cannot do without the concept of nature no more than without the concept of freedom.
>
> I. Kant, *Grundlegung zur Metaphysik der Sitten* (1785)

1 Premises and Introduction

In the Editors' original words the present book was "intended to contribute to creating a counterpoise to reductionism and other 'isms', each of which being separately taken by many to be a rationally unavoidable consequence of physics for the philosophy of mind. In contrast to the old physics, quantum physics is not only empirically correct, but also promises, in its *indeterminism* and *holism*, to have more to oppose to reductionism and other 'isms' than the arguments of traditional psycho-physical dualism could offer".

While being fully sympathetic with a counterpoise to physicalistic reductionism, I do not share the view according to which quantum theory (QT), in its current form, could provide effective weapons to such counterpoise in a substantially different way from classical physics (CP). I shall defend the view according to which more efficacious weapons for this basic philosophical struggle could derive from a new approach to the perennial unsolved issue of Free-Will in a form pointing to a non-traditional perspective about *psycho-physical dualism.*

As John Earman observed, "the determinism-(indeterminism)/Free-Will controversy has all the earmarks of a dead problem and no advances in philosophy of science or cognitive psychology seem to move the problem forward". While agreeing with this point of view, I believe that the main responsibility for the deadly aspect of the issue is to be found in the harsh and rigid dichotomy *determinism / indeterminism*[1] only, in which Free-Will has been traditionally entrapped.

---

[1] Put simply, determinism usually refers to (yet not in this Essay) the doctrine according to which the laws of nature alone are sufficient to determine the whole history of the world if a

In this Essay I shall propose a view that bypasses the strictures of such a traditional clash without any recourse to the so-called indeterminism of QT. I do agree that the specific "indeterminism" deriving by the *atomization* of *action* and the consequent spatiotemporal limitations embodied in the *superposition principle* may play some very important role in clarifying many aspects of the neural *functioning* of the brain. Even more, I believe that new suggestions are possible for both the philosophy of quantum theory and the philosophy of mind.[2] Yet *chance* by itself is not enough for *Freedom*. Indeed, *stochasticity* could at most somehow reproduce in a new sophisticated fashion the classical atomic *swerve* of Epicurus and Lucretius. Some additional *non-physical agency*, having the ontological capacity of 1) freely *modifying* the quantum *probability distribution* and 2) freely *making a choice for a single outcome*, would be required anyway,[3] pointing—in one way or another—to some refined kind of epistemic *psycho-physical dualism*.[4] In this perspective, I am even willing to concede that the intimate interplay between micro-neural structure and potentiality of quantum actualizations may provide a *locus* where a *non-physical causation* could be exerted. This, of course would be relevant to the Mind-Body issue. Yet, a problematic *dis-solution* of this conundrum is not necessary to understand Free Will. If the problem of the *compatibility* between a *causality through Freedom*[5] (or *non-physical causality*) and the *necessity* of physical laws could not be dealt with and solved within the perimeter of *macroscopic* classical physics[6] (where a solution is required locally without the pressing need of retracing it back to the Mind-Body stage), I seriously wonder whether QT could successfully come to the rescue.[7] In conclusion, I think

---

complete initial stretch of that history is given.

2 In Pauri 2011 I expounded a proposal concerning the possible *role* of the *quantum neural structure* of the brain in the *macro-objectification* issue of QT. This proposal leads to a form of *naturalized transcendentalism* which might go in the direction of a counterpoise to reductionism.

3 The claim that indeterminism *is not enough* is true even regarding the so-called *Two-Stage Models for Free Will*, which include quantum indeterminism as a first stage. In my opinion such Models fail to solve the problem, by simply shifting it to the second stage. See, e.g. Heisenberg 2009, 164–165.

4 I will not take a stand here about the ontological issue of dualism of substances in Meixner's sense. See Meixner 2004.

5 See, e.g., Kant 1963 II, p. 476.

6 Both relativistic and non-relativistic.

7 I insist that the problem must be solved at the level of physics. Among other things, I do not believe that current biology, as a global scientific image of living organisms, has any gap—aside quantum indeterminism—in which to search for the compatibility of Free-Will. Incidentally, I observe that state-of-the-art biology lacks even the very theoretical definition of "organism"; see Longo, and Montévil 2013. An "organism" has a 'unity *per sé*' which cannot simply be seen as *emerging from below*.

that Popper's attempt[8] to solve the problem of Freedom by simply rejecting the thesis of the so-called *causal closure* of the physical world is radically inadequate.

My starting point is that freedom of the *Will* (*Free-will*) and related issues as *Freedom of Choice*, *Freedom of Action* in the 'physical world' and, above all, the capacity of discriminating *truthfulness* in rational argumentation (which I call 'Freedom of Meaningfulness') are *ontologically* real features of the World. I claim that freedom of the Will, in any valuable sense of the term, is the *power* of living organisms to be the *ultimate originators* and sustainers of their own 'purposes' and 'wills', so that the causal or explanatory chains of actions which, in their essence, must be traced back to the wills *must* come to an end in such wills. Moreover, I strongly maintain that if we were not able to *freely will* and *freely act* in the world, we could not even be able to *freely believe* and to be free to *meaningfully think*, and *vice versa*. On the other hand, we are, indeed, *condemned* to *meaningfulness*. Terms such as *consciousness* (though not *self-consciousness*), *awareness* or *sentience* are here used as synonyms of the general term *subjectivity*. I am not supposing, of course, that the same richness of properties could be found at every level of evolutionary development, but I would maintain that some basic elements of subjectivity extend—with proper gradation—all the way down to all *living organisms*. Finally, I declare from the start that I accept the *reality* of *potentiality* (and the connected *finality*) and its *actualizations* as a fundamental feature of the *living beings* and, therefore, of the *world*. *Causality through Freedom* and *Finality* are two aspects of the same fact. The *bi-unity* of causality and finality is the very essence of the *future-past relation* and of the agents' freedom.

I wish to start by highlighting the possible cooperation of two distinct conditions proper to the physical inquiry which, so far, have not been put in active connection and have therefore proceeded quite separately. Namely: (a) the fact that "the whole of our philosophy of science is based on the assumption that one is *free* to perform any experiment";[9] (b) the fact that, *by their very structure*, physical laws are entirely *silent* on the *values* and *chronological locations* of *initial conditions*.

I shall then attempt to give: (a') philosophical substance to the first condition by confronting it with the threats of *aporias* if interpreted as a mere pragmatic, purely programmatic utterance which is taken for granted in the daily practice and epistemic purposes of laboratory physics, and (b') exploit the philosophical and scientific consequences of (b) for Free-Will.

With this in view, the argument will be centered on a revision and new development of the *libertarian* Kantian statement quoted in the *epigraph*, taken

---

8 See Popper, and Eccles 1997.
9 See Hawking, and Ellis 1973, 189.

as unavoidable to shun *aporias*.[10] More precisely, my analysis will be grounded on the following distinct, yet coherent and converging, lines of thought: (i) The introduction of what I call "ontological-phenomenological *aporia*" (OPA), which stresses the *aporetic* nature of any alleged basic *distinction* between *reality* and (*universal*) *illusion*, when claimed *from the point of view* of any physicalistic ontological reductionism; (ii) The instrumentally provisional (in view of the subsequent discussion) formulation of a *Postulate of Freedom*, connected to a causality through Freedom (C_{free}). (iii) The critical re-visitation—a fundamental step in my argument[11]—of the structural limitations of the Galilean methodological foundations of (local) physics, viz. the programmatic *blockade* of the *essential* individuation of entities with their consequent *abstract objectification* and *mathematization*, conjoined with the operative role of *theoretically planned* and *repeatable* experiments. The Galilean *blockade* leads to the characterization of a new *reduced ontology* (RO)[12] for the physical theories. The RO allows to speak of the so-called *causal closure* of physics [with respect to the *physical causality* (C_{ph})], as well as of *determinism* or *indeterminism* of a '*theory*', within the scientific image provided by the *physical description of the world*. Then, my basic claim will be that beyond the typical *idealizations* of the RO and physical *theories*, there is an *open world* (OW), which is neither (*nomologically*) *deterministic* nor *indeterministic*. OW should be better characterized as *undetermined*, not in a nomological sense, rather in the *meta-physical* sense of allowing for *ontological potentialities* (from here on, the term "meta-physical"—with lower case "m"—refers to OW, just in the sense of staying *beyond* RO). It is important to stress from the beginning that the causal "openness" of OW does not contradict the *local validity* of the RO and the physical laws ruling it.

All this will exhaust the task of giving substance to the conditions listed under the above (a). Finally, crucial to my main argument [and fulfilling the requirements of (b') above] is the fact that the physical laws, ensuing the Galilean *blockade*, be (as indeed, theoretically, they are) totally *silent* on the *values* and *chronological locations* of *initial conditions*, which must be carefully *chosen* in *planning* any laboratory *experiment*.

---

10 See Sections 2 and 8.1. Of course, I am not going to be committed to the whole structure of the Kantian *a-priori*.
11 In Pauri 2008, it is argued how this revisitation may also have a critical relevance for the history of modern philosophy and the contemporary philosophical sentiment at large.
12 As a first example, the ontology of CP is an ontology of "mass-points" (Euler), a concept which is a true prototype of the synthesis *of simple localization* (Whitehead) and *vis insita* of inertia.

Therefore, the structure of the Galilean foundations—on pain of OPA for the very concept of physical experiment *qua* experiment—entails *necessity* and *sufficiency* for freedom concerning the choices of *initial conditions*, thus validating the Postulate of Freedom, at least in its action component. This means that, unlike the physical description in terms of the RO (which includes the constitution of *physical time*), the OW—bypassing the crucial clash of *determinism* vs. *indeterminism*—allows for the *causality through freedom* (C_{free})[13] and the reality of *free actions* operated by living organisms. Thus, we can conclude that the reality of (C_{free}) is a *necessary* philosophical prerequisite for the very Galilean foundation of *local* Physics and, in a deeper sense, a *'transcendental' condition of the experimental method*. This fulfills task (b) above, and shows a direct connection between the structure of physical laws and the mind-body issue.

Coming back to OW, let me remark that, unless one *aporetically* maintains (in the sense of OPA) that the very *freedom* of choosing the initial conditions is *illusory*, it follows that the *real* world must be, as already stressed, *macroscopically undetermined* locally to such an extent to provide enough freedom for the experimentalist's *choices*. Note that it would not do to define OW as *undetermined simpliciter*, since *after* a free action, OW *must* be FAPP[14] *law like* and deterministic. Thus, as said above, OW allows *actualization of potentialities*,[15] so that, in my view, the free agent's role is not simply that of *narrowing down* the number of nomologically possible futures. As we will see, this leads to a conception of OW as a world which cannot be *presently* ruled by *a thorough-going system of lawful connections* between events. In conclusion, Free-Will is inexorably a problem of *cosmological order* (in a Kantian sense).

It is then surprising to discover that Kant himself overlooked some fundamental consequences of the causality through freedom which he strongly upheld, as regards just the *universality* and *thorough-going* cosmological validity of a system of lawful connections among events. The consequences of Kant's oversight will be illustrated by showing the implications of the *tenseless* vs. *tense* temporal *nexus* of the philosophy of Time concerning Free-Will, with particular emphasis on the structural limitations of the *(tenseless)* description of the *actualized events* in Minkowski's spacetime [i.e., within the RO of the special

---

13 As distinct from the *physical causality* (C_{ph}) which is effective within the RO. See Section 5.
14 For All Practical Purposes.
15 Let me recall that, in the quantum case, the lack of *nomological cause* for the actualizations leads to the notion of Heisenberg-Shimony's "potentiality"; see Shimony 1993a, 140. Such concept of potentiality could also be seen as a *reduction* (within RO) of the free agent's potentiality typical of OW.

theory of relativity (STR)]. It will then be clear that a *literal interpretation*[16] of the RO is just another manifestation of OPA.

The criticism of such ontological implications leads in turn to what I call the *real temporality* of the OW revealed by the free actions of living organisms. Precisely, by exploiting the outcomes of Kant's (seemingly unknown) oversight in the discussion of the Third Antinomy of Reason, the reality of C_{free} appears to lead to an *additional, tensed* and *irreversible*, temporal structure of the OW with respect to the standard *tenseless structure* of physical time. The *footprints* of the (*tensed*) *Real Time* of OW disclose themselves, within the relativistic RO, as *irreversible (tenseless)* traces generated by the *local free actions* of living subjects. Such traces (like any other physical process) are, necessarily, *causally* ordered in agreement with the standard relativistic causality (C_{ph}). In this way, an *arrow* of time is projected, so to speak, into the *physical description of the world*, by the structure of phenomenological time. Assuming that the *before/after* experience is a fundamental ontological property of all living organisms, the projected *phenomenological arrow* appears as an additional ontological structure coherently superimposed on the *physical arrow* of time. This is not a *physical explanation* of the latter, of course, but since *organisms* belong to the world, the *reality* of freedom of action turns out to be strictly entangled not only with an ontological bearing of the *anisotropic phenomenological time,* but also with the physical *arrow of time,* a fundamental *de facto* property of the *physical description* which, notoriously, has never been theoretically explained. Within this perspective, *physical time* is just a *reduction* (in the sense of the RO) of *Real Time*, like C_{ph} is a *reduction* of C_{free}. In conclusion: (i) my argumentation bypasses the strictures of both *compatibilism* and *incompatibilism* which are usually invoked against Free-Will, and that, accordingly, (ii) there is no lack of fit of Free-Will with the modern scientific image of the world; and finally, even more, that (iii) the reality of C_{free} is a '*transcendental*' *condition*[17] of the *experimental method* founded by Galilei.

In this Essay the argument shall be developed throughout according to a principle of metaphysical *parsimony,* with no suggestion of *positive* Metaphysics (M). Correspondingly, the overall logical structure is that of a *reductio ad absurdum*, by which I try to individuate a perspective avoiding deeply antinomic conclusions.

---

[16] In this Essay, literal *interpretation* of the RO refers to interpreting (*aporetically*) the RO as *ontology tout court* of the knowable world.
[17] Surely not in the strict Kantian sense of the term. See, e.g., Pauri 2011 for a notion of *naturalized transcendentalism*.

2 The Ontological-Phenomenological Aporia[18]

I state that a philosophical argument concerning the knowledge of reality falls into OPA if, in attributing a fundamental primacy to the issue of 'meaningfulness', it: 1) explicitly or implicitly entails the necessity of maintaining a *meaningful* distinction between *reality* and *appearances*, or even (universal) *illusions*;[19] however: 2) It is based upon an explicit or implicit *ontology* that does not have the power of explaining the consequent *phenomenology* in an acceptable *meaningful* way. In particular, it does not have the power of accounting for possible phenomenological and logical contradictions, including the very *distinction* between *reality* and alleged *appearances* (or universal *illusions*), *qua* appearances.

A case study of an (ontological) world view which is *aporetic* in the sense of OPA is Steven Weinberg's "Deterministic Grand Reductionism".[20] Appendix A and Fig. II show how Weinberg's explicit physicalistic ontology cannot account for the phenomenology consisting of two allegedly (C_{ph})-causally generated meaningful utterances, in which opposed or contradictory theses are maintained, e.g. by Steven Weinberg himself and his friend George. Substantially, no sufficient reason exists for a belief or sentence to be true or false (or *illusory*). Even more—within Weinberg's declared ontology (as in any physicalistic and deterministic world-view)—no conceptual room exists, or can be added, for any sensible notion of meaning and truth validating Weinberg's or anyone else's utterances, except for the rough fact of the world that what happens, happens as it happens, like a storm or an earthquake. Another significant case of OPA is exhibited by Hilary Putnam's interpretation of Minkowski's spacetime in the Special Theory of Relativity (see Section 6.2).

---

[18] OPA is a generalization of what Abner Shimony calls *Phenomenological Principle*. See Shimony 1993b, 278–279.
[19] The standard case of "the broken stiff in the water" is a *non-universal* illusion which can be explained as trivial phenomenology grounded on the ontology of physical theories. The alleged "qualia eliminativism" (see Dennett 1979, 2003), i.e. the interpretation of "qualia" as "universal phenomenological illusions" would require some *explicit* scientific (ontological) explanation of our fundamental experience of "qualia" themselves. It is up to the reader to check whether "qualia eliminativism" has offered such explicit explanation or is, instead, on pain of OPA. Even more, Dennett claims he has succeeded in a radical elimination of any kind of *phenomenology* and consequently could maintain to have radically circumvented both the 'grain objection' and OPA, in a single blow. The same is true for the attribution of 'universal illusoriness' to the fundamental *transubjective* and *permanent* experience of 'nowness'.
[20] See Weinberg 1995.

3 Self determination and Agency in the world. The Causality through Freedom: a first look

I now state a (provisional) *Postulate of Freedom*: "By exploiting the very structure of the physical laws—viz., in particular, the crucial fact that such laws are *totally silent* on *initial conditions*—we can freely intervene in the world as described by physics. More specifically, within certain limits, we have the power of: (i) determining *our own purposes* in the sense of being the *ultimate originators* and sustainers of them. This means, in particular, that we could *do otherwise* in *exactly the same circumstances*, provided the latter be truly *circum-stances* i.e., *ontologically 'outside'* our core *subject*;[21] (ii) giving rise *freely* and *selectively* to a *local* physical state of affairs such that, as *efficient* cause and exploiting coherently our *knowledge of physics* (viz., e.g., the relativistic causality C_{ph} of the RO, see later), with a great probability it later brings about the effect we have *freely chosen* to realize. This is a provisional postulate of a Causality through Freedom C_{free} (distinct from C_{ph}), which *actualizes* real *potentialities* in the world".

4 Revisiting Galilei's foundations of physics: the fundamental *blockade*, the *Reduced Ontology*, and the *Open World*

Galilei's well-known intuitions were implemented, through a radical philosophical transformation taking place during the Renaissance, by means of a *blockade* of the process of *essential individuation* of entities. 'Accidental' and 'essential' properties were no longer taken into account for the understanding of *temporal change* and *transformations* of individual entities. The *blockade* then leads to the replacement of the *concreteness* and *quality* of experience with the *abstractness* and *idealization* exploitable by *mathematization*. The transformation of the world of experienced objects into a world of abstract mathematized entities entails the simultaneous and implicit ablation of the experiencing *subject* from the field of entities that physics is entitled to understand, *at least until proved to the contrary*.

---

[21] Clearly point (i) would be self-contradictory due to the *indiscernibility* principle if "identity" were predicated of all circumstances and conditions, including those belonging to the *ontological essence of subjectivity*, such as the *final causes* (which are the true causes of the choice of action).

The blockade and its developments have been made possible by exploiting two basic *facts* of the *actual structure* of the external world. Namely, under suitable approximations (or FAPP): a) The *factual* possibility of a *local physics* which allows to deal with *isolated physical systems or objects*. b) The *factual* existence in the Universe of autonomous sub-totalities within which a *stability of recurrences* can be acknowledged.[22] Starting from *phenomenological time,* this leads to the constitution of *physical time* as a *primary reduction* which, according to Whitehead, bifurcates nature. Physical time is *relational* by constitution (see later).

Another *founding idealization* is the separation of the world into two (or three) parts: (i) the *object* to be defined as an *isolated physical system*; (ii) the *observing, theorizing* and *acting subject*, who often takes on the ambiguous role of 'observer' or even 'measuring apparatus'; (iii) the *remnant of the world* (which may include the subject). The variability of this last part, initially taken as irrelevant to the immediate scope of defining the physical system, determines the *irreducibly contingent* component of the physical description in the form of *initial* conditions. This partition allows for the determination of *possible* temporal *evolutions* as *physical laws*,[23] in connection with *different choices* of the *initial conditions*.

---

[22] Let us assume for simplicity that our accessible region of space-time (or of any other characteristic region in which experiments must be performed) is describable, in the scientific image, by, e.g., a Hamiltonian system. The couple of facts above would imply that the possible positive *Lyapunov exponents* be sufficiently small or can be rendered such by the experimenter, in such a way that the accessible region, even if *chaotic*, would appear to be so only after a reasonable time span. Examples of this condition may be an electromagnetic or thermal isolation of the system to be studied in a laboratory scenario. In the case of astrophysical observations, including the solar system, the positive *Lyapunov exponents* are *already* so small that the relative time span can be supposed to be of the order of magnitude of the age of the universe. A negative case is that of meteorology in which a typical time free of chaoticity does not exceed three days. It should be sufficiently clear that I am interested here mainly in a question of principle, just in view of the Postulate of Freedom.

[23] For the sake of brevity I have limited myself to a classical *laboratory scenario*, as a case study. Admittedly, typical Galilean conditions such as the indefinite *temporal repeatability* of the relevant relations with the remnant of the world, or the *irrelevance of the spatial relations* between a suitable region including the physical system and the remnant of the world (*spatiotemporal homogeneity*), are no longer simply satisfied in the case of general relativity and must be adapted. The situation of astronomy and astrophysics is still recoverable due to the *multiplicity* of similar physical systems under scrutiny, a fact that mimics the choice of initial conditions, while the basic conditions are not met in the case of *cosmology*. Indeed, I believe that "the Universe as a whole cannot be considered as a scientific object in any sense that such words have had in the historical development of physics"; see Pauri 1991, 2001 and a related comment in Torretti 2000. Note that, unlike the great theories of standard physics, cosmological theories, concerning in particular hypothetical structures lying beyond the reach of 'direct'

In this way, the *blockade* of *essential individuation* leads to a peculiar *new individuation* or, better, *specification* of entities constituting a new *reduced abstract ontology* (RO). The important point is that, within the limits of the internal methodological approximations of the physical description, *this specification* is considered *exhaustive* or "complete". This allows to maintain—in the new sense—terms such as *the ontology of a theory*. Typical instances are ontologies of: *mass points, classical fields, elementary particles, antimatter, identical particles, relativistic quantum fields, metric fields*. Other instances are: the notion of 'observer' in Lange's sense (e.g., an observer represented by an infinite reference frame[24]), up to the very concept of *physical time*, and so on.[25] Once the RO is established, any classical physical action is described in space-time as an exchange of 4-momentum among the interacting parts of a whole.

It is just in force of the *exhaustiveness* or *completeness* of the RO relatively to a given *theory* (which is *basic* for the mathematical concept of "initial conditions") that one can speak unambiguously of *determinism* or *indeterminism* of a '*theory*', while from the premises of the RO it follows that no one is allowed to attribute *determinism* (or *indeterminism* either) to the world *simpliciter*. Aside from the idealized *reduction*, there is in fact no *natural minimality* in the *concrete multiplicity* of the real world upon which to base both the dynamical description and a consistent *concept of initial conditions*.[26] Therefore, *absence of determinism* (in the world) is not equivalent to *indeterminism* in the ordinary sense of *chance*. It is important to stress here that the meaning of the concepts *determinism/indeterminism* we are

---

astrophysical observations and mixing conflicting theories like QT and General Relativity (GR), lack a *domain of validity*, a fact that makes cosmology *epistemically unstable* in front of possible sudden astrophysical discoveries. The Galilean-like description is of course even more inadequate for *quantum* phenomena and requires important adaptations.

24 Or something more complex and refined as in GR. See Lusanna, and Pauri 2007.

25 A very particular case is the ontology of quantum theory which, according to Quine, is "an *ontology of abstract entities* (though not *mental ones*)". The fundamental point is that, in this case, '*abstract*' assumes the stronger qualification of '*non spatiotemporal*' so that such 'entities' can be represented only *symbolically*.

26 Consider the evasiveness of the concept of Laplace demon, with reference to "des êtres qui la composent" in the well-known Laplace quotation: "Une Intelligence qui, pour un moment donné, connaîtrait, toutes les forces dont la nature est animée, et la situation respective *des êtres qui la composent*, si d'ailleurs elle était assez vaste pour soumettre ces données à l'analyse, embrasserait dans la même formule les mouvements des plus grands corps de l'univers et ceux du plus léger atome; rien ne serait incertain pur elle, et l'avenir, comme le passé, serait présent à ses yeux"; see Laplace 1812. Furthermore, consider the following important statement: "The ideas of determinism, probability, indeterminism, and reduction can be significantly discussed only if they are directed to the *theories or formulations of a science*, and not to its *subject matter*" (obviously *reduction* only in the weak or methodological sense); see Nagel, E. 1939.

talking about for the specific issue of Free Will must just be the *scientific* ones and not the *generic terms* of the philosophical tradition. Not only does the primary ontological reduction leave out of the picture the experiencing subjects with their *qualia* but, more generally, a whole background of *happenings* of the world which could not be mathematized in principle. In other words, *beyond* the typical idealizations of the *physical description of the world* there is an Open World (OW). Unlike the RO, which is *by construction C_{ph}-causally closed*, the Open World allows for *free causality chains* or *actualization of potentialities* so that it could be better defined as *undetermined*. In conclusion, *beyond* the description provided by the RO, there is certainly *something out there*, independently of our representations. Yet, it is an 'open world' and there is no 'physical world' *simpliciter*, let alone a *physically closed* world. In conclusion, we can only talk of a *physical description of the world* (in terms of RO).

Let me come now to the *fundamental feature* of the RO indicated above: namely that the physical laws are *entirely silent* about the *values* and *chronological locations*—i.e., the *actual fixing*—of *the initial conditions*. In fact, *within a laboratory scenario*, while physical laws rule *possible* time evolutions, it is only after the *initial conditions* are selected and *fixed* that we get a (*theoretical model* of) an *actual* time evolution. In other words, although the technical concept of *initial conditions* is an element of the RO, practically fixing or modifying them is always a true *actualization of a potentiality* in the real world. I claim, accordingly, that the world, as *describable* by *classic physics*, contains *sufficient* and *necessary* reasons for Freedom. For (as living and sentient beings), we can—factually—fix or modify existing *initial conditions*, which *are not pre-selected by physical laws*. In this sense *sufficiency* holds. On the other hand, the condition of freedom is *necessary* since if the initial conditions could not be *freely fixed* or modified any experiment, *qua* experiment (recall Galilei's "sensate esperienze"), would be *meaningless* in view of its methodological role and *aporetic* under OPA.²⁷ In conclusion, we act in the world according to the Freedom Postulate. Finally, I claim that living beings, *qua* living and feeling, cannot be considered as *physical systems* in the sense of the RO (except under limited perspectives).²⁸ In particular, I hold that the notion of *initial* conditions *at a fixed physical time* cannot be applied to them. As Whitehead pointed out, *there is no life at an instant*.²⁹

---

27 Note that the Freedom of choice of the initial conditions is *even more necessary* in QT. See, e.g., Conway, and Kochen 2006, 1441.
28 Physical laws have been tested on living organisms, so to speak, only *sectionally* on functions previously isolated. Surely, the possibility of considering them globally is not precluded, within the RO of a larger system, as open *thermodynamical systems*. Also, nothing prevents from treating living organisms by idealizing them within a form of RO which does not take their *individuality* into consideration.
29 "In biology the concept of organism cannot be expressed in terms of 'matter distribution at a

It is just the a-priori indescribable variability (within the RO) of the relations with the *remnant of the world*, that leaves *room* for the freedom of the acting subject to fix the initial conditions. As a matter of fact, any free action establishing the values of initial conditions necessarily generates, besides the desired effect, additional incalculable variations on *all* the relativistic C_{ph}-causal chains. This again enlightens the cosmological connection of the reality of Freedom.[30]

A source of wonder (yet, after the disgregation of the Kantian *a-priori*, of inexplicableness[31]) is the fact that the RO obtained by the *blockade* be able to disclose in mathematical form an extraordinary, sophisticated and stable *relational structure*, even beyond the immediately perceptible reality. However, only the extreme complexity of the relational texture can give the fallacious impression that, instead of the *how*, the *why* or better *the quid* of things and processes be explained (a fact that the Galilean *blockade* structurally forbids). What said above limits the applicability conditions of the RO and—consequently—even the domain of meaning of the *physical description of the world*. For these reasons the wonderful, yet enigmatic success of the mathematization proper of the RO *does not license anybody* to attempt a *Carnap-type reconstruction of the open world* starting from the *physical description*, i.e. taking the RO as ontology *tout court*. This interpretation of the RO unavoidably leads to OPA.

A final remark must be issued. Obviously, physics is not limited to the *theoretical* description of the world. It includes the actual, *unavoidably tensed*, experimental world of living physicists in OW, with their *physical intui*tion, and the *correspondence rules* between theoretical terms and empirical operations.

5 Self determination and Agency in the world: a second look

Let me make a few comments on the issue of Causation. Many critics, compatibilists and libertarians too,[32] would say that, in relation to the *reasons/causes dichotomy*, my concept of Freedom is of the 'agent-causation', or 'non-occurrent' (*bad*) kind, in the sense that I admit the possibility of *direct causation of events* by

---

time t'. The essence of an organism is that of being an entity that *functions* in space. Now, functioning requires a *temporal duration*. Something as 'life at an instant' cannot exist. Life is too *obstinately concrete* for being *localized within an extended element of instantaneous space*"; see Whitehead 1955, italics mine.

30 For very explicit statements in this regards see Bohm 1980 and Nagel, T. 1987.
31 See Wigner 1960.
32 See Kane 1996, 120–123; 190–192.

an *agent* (viz., in current emergentist terms, a *downward* causation) as opposed to the current notion based on the *event-event* relation³³. This, obviously, is true. The critics of such a *non-occurrent* causation object that it is *unintelligible* and *empty* of *explanatory power*. Certainly, it is *unintelligible and empty*, in physical terms, i.e. *relations among events* (no more, however than *consciousness* itself or the *mind-body* problem). Yet, this *unintelligibility* is in my view an asset rather than a liability. Actually, C_{free} must be *unintelligible* in terms of elements of RO if it has to be *free*. Furthermore, I am not claiming that an agent is free if his Free-Will does not depend on antecedents, but only if it does not follow from *physically describable* antecedents nor, *necessarily*, from other antecedents which do not include *final causes*. Indeed, it will certainly depend upon antecedents such as meaningful *reasons*, memories, *meaningful information*,³⁴ even *qualia*, and so on, yet *non-necessarily*, and without physical constraints in the sense of the *physical description of the world* (i.e. elements of the RO). On the other hand, the critics' ascription of *unintelligibity* of the agent-causation presupposes an implicit imagined 'resolution' of the mind-body *nexus*, available in the near future in a reductionist perspective.³⁵ Now, consider the

---

33 According to the event-event view, we have, for example, in a reductionist vein: "The cat (agent!) caused the lamp to fall (event)" is *elliptical* for "The cat's jumping on the table (event!) caused the lamp to fall (event)"; see Kane 1996, 120. Of course this is a meaningful description of a *physical process*, yet it does not explain why the cat jumped on the table. It seems to me, however, that, once *extracted* from the network of *temporally ordered physical causal chains* ruled by physical laws, the direct causation *event-event* is even more *unintelligible* than the agent-event causation, given also the fact that the concept of *event* is a highly idealized notion that acquires a clear role only within RO. This in some way matches what Kant claims in the discussion of the Third Antinomy of reason: "Since the power of spontaneously beginning a series in time is thereby proved *though not understood* […] How such power is possible is not a question which requires to be answered in this case, *any more than in regard to causality in accordance with the laws of nature*". For "*we are in the least able to comprehend how it can be possible that through one existence, the existence of another is determined*"; see Kant 1787, italics mine. If, for the sake of argument, we instrumentally replace the Kantian a-priori with the negative power of OPA in order to ground the causality through freedom, the weakness of the alleged intelligibility of the event-event causation is manifest. After all, Kant accepted Hume's analysis of the causal relations almost entirely, as relation between *events*. However, our causal verbs normally take persons or *things as subjects* and imply a *meta-physical* notion of cause that reduces to the event-event C_{ph} within the RO.
34 I fully agree with Meixner, according to whom a piece of 'information' (in its general meaning, not in the specific sense of *information theory*!) which a subject becomes aware of, "has, as it were, an address written on it" (*for*ness). See Meixner 2004, 317. In my words, *meaningful* information is *information for someone*, even when it is embodied in an analogical support like, e.g., an electromagnetic wave: then the subject elaborates the 'information' in its general meaning, not its digital codification, or the electromagnetic wave.
35 I wishfully recall a sentence of the great neuroscientist Charles R. Sherrington who, in front

usual deterministic-compatibilist position that goes back to Hobbes ("The agents are self-determining *only* in the sense that nothing prevents them from doing what they *want*", since "obviously we *deceive ourselves* in pretending to be free in determining *what we want*").[36] However, if we are (*universally*) *deceived* to think that "we are also free to want what we want, independently of empirical antecedents", why should we (and Hobbes with us) not be (*universally*) *deceived* also in thinking that "we are free to think *truthfully* what we think", independently of empirical antecedents? I claim that Hobbes and his followers are just *aporetic* under OPA.

Finally, I would like to add some suitable specifications about the concept of *physical causation*. Firstly, the extent to which the very concept of *causation* is meaningful in physics is not crystal clear. This concept is originally a *meta-physical* concept referred to OW. Then it is 'exploited physically' at the level of the RO[37] where, within a definite *theory*, it reduces, essentially, to a network of *mathematical relations* within a (classical or relativistic) *tenseless temporal* (and *topological*) *order* (see later). On the other hand, this reduction of causation is precisely nothing more than what is needed for physics, for in this way this concept boils down in fact to that of *deterministic lawful connection* between time-ordered events. Such a reduction of the concept of physical causality to that of *determinism* appears to be even more revealing if one considers the basic *ontological* nature of the *probabilities* involved in the irreversible macroscopic *actualizations* of *quantum superpositions*: just *absence of determinism* is tantamount to *absence of causes*! Furthermore, this clarification at the level of RO casts some light on the concept of the *causal closure of physics* itself. It is no wonder that such *causal closure* of physics is typically invoked to deny causal powers to mental properties, as it is no wonder that physicalists question-beggingly assert the existence of an unavoidable conflict with the physical conservation laws.[38] Clearly, the C_{ph}-*causal closure* of physics is only valid under the condition that the *theorizing and acting subject* does not intervene in the *dynamic play* of the game *once the initial conditions are settled* or until

---

of any possible and alleged 'explanation' of consciousness, declared: "In *that day*, I will be happy to understand what such *explanation* could *mean* in the first place"; see Sherrington 1937-1938. In my view, scientific *explanation* means a *relational* explication which connects *different regions* of the domain of *objectivity* in terms of physical laws and principles. Thus, I too, on earth, would like to understand what it could mean to *objectively explicate* a jump from the domain of objectivity to that of *subjectivity*, if the latter is not already surreptitiously put in.
36 Italics mine.
37 Note that this is true even regarding the concept of '*force*'. Newton was perfectly aware of the metaphysical residue in his concept of force. Only with Enlightenment—viz. with Laplace—Newtonian mechanics was interpreted as a fully 'mechanistic' theory.
38 See also Meixner 2006, § 7 "The Insect-objection".

when the play is possibly changed by an acting experimenter who, again freely, wants to do so. Even more, as Uwe Meixner pointedly stressed,[39] "The *principles* of causal closure of the *physical world*—constantly invoked against the nonphysical causation of the physical—are neither principles of the logic of causation, nor principles of physics, but postulates of materialist metaphysics". From my point of view, what they are is very clear: untutored consequences of applying Carnap's *dogma*[40] to the interpretation of the RO as *ontology tout court*, i.e. a deformation which lies at the origin of almost all the *aporias* of materialist *metaphysical* statements. In conclusion, the working of the causality through freedom is not more unintelligible than that of physical causation, once the effectiveness of *final causes* is excluded.

6 Time and Free-Will

The issue of Free-Will is strictly connected with the form according to which *subjectivity stays in time*. Thus philosophy of time is directly entangled in the inner core of Free-Will and Mind-Body through the *nexus time-causality*. I shall assume that the reader is sufficiently acquainted with the lasting 20th century debate about *language* and *ontology* of time. A brief sketch of the premises about Language is given in Appendix B, where a basic distinction is made between *tense* (A) and *tenseless* (B) temporal determinations. The crucial point for us is the following *ontological alternative*. B-Theory: no *real demarcation* exists between past and future, so that no *objective transient* properties or facts exist in the world (future is *logically closed* as *is the past,* and also *causally closed* if the world is *deterministic*). *Nowness* is *unreal* in the sense of being *only* Mind-dependent and non-objective (even more, in its consequences for actions, 'illusory'). Real events are ordered by B-series relations within the set (call it *W*) of 'all real events of the world' (Plato does really exist in the sense that—unlike the *hippogryph*—he belongs to W). A-Theory: Every subject lives *constantly in his own unique and eternal now*, and maintains his personal identity in the experienced (so-called) 'flow' of time. A *real ontological threshold* exists between merely *possible* and *actualized* events (future is *logically and C_{ph}-causally open*, see later). *Nowness* and *transiency* are ontologically *real* and the 'now' is when *free actions actualize possible events*. We know, on the other hand, that from an *operational* relativistic

---
39 See Meixner 2008, italics mine.
40 The view according to which philosophical reflection must only start from both logical structures and scientific acquisitions (of physics, in particular). This clearly entails bypassing any *critical assessment* of the *very foundations* of physics itself. See Carnap 1928.

standpoint, concerning spatiotemporal *measures,* the possible sharing of the *now* by many observers is only *local*. I claim that the notion of *nowness* is *meta-physical* in the sense adopted in this Essay, i.e. in the sense of being a property of OW.

6.1 Physical time

A fundamental component of the RO is *physical time*. Its constitution is a sort of *primary idealization* which, *abstracting* from the *(tensed) phenomenological time*, introduces a basic bifurcation of nature. The *possibility* of this constitution is guaranteed by *factual* condition (b) (see Section 4) exploited by the Galilean *blockade*. Unlike *phenomenological* time, *physical time* is *relational* by constitution and all of its determinations belong to some *B-series*. This entails from the beginning that physical laws are intended to be *valid for any time* and cannot therefore *single out a particular instant* of time as 'the now'. Likewise, no physical experiment can discern whether a 'particular time', shown by the hand of a clock, is 'present' or not. Unlike the laboratory *actual practice* of the experimentalist, the theoretical structure of physics refuses any kind of 'nowness'.[41]

6.2 The *Reduced Ontology* of the Special Theory of Relativity and its Interpretation[42]

A basic tool belonging to the RO of STR is the four-dimensional Minkowski spacetime M^4, in which all kinds of spatiotemporal measures can be settled. It is a space of 'unchanging' events (E)[43] and it also provides a realization of the set W[44] of all real *physical events* of the world which are tenselessly ordered by B-series relations, so to speak, once for all. M^4 is, in addition, endowed by a fundamental

---

[41] More on my views about physical time, concerning in particular the striking temporal efficacy of the Galilean *blockade*, can be found in Pauri 1991, 1997a, 1997b, 1998, 2001, 2008.
[42] From here on I shall discuss the main temporal features of Free-Will within the framework of the STR, only because of its stronger time-causality relations. I could have dealt with the issue within CP as well.
[43] The concept of event embodies the notion of *spatiotemporally-objective* and *simple localization* of physical properties. While things change in time, *events* are *tenselessly defined* by their *attributes* and *do not change*. They either take place or they do not. Note that there are no *things* in M^4, only *events*, *world-lines* (or "spatiotemporal localized histories"), and *chunks of world-lines* (or "histories of extended bodies").
[44] See also Appendix B.

(C_{ph})-*causal structure* of "light cones"[45] [a light cone (LC), with vertex at any point-event E, is defined by all events of *ingoing* paths of light (*backward* LC) and *outgoing* (*forward* LC) at E, and by all the *internal* point-events. LC defines the spatiotemporal zones of C_{ph}-*causal influence*, respectively *passive* (*backwards* LC) and *active* (*forward* LC) of E]. Thus, from such structure of light cones and the consequences of Einstein postulates concerning the conventionality of *distant simultaneity*, it follows that: (i) The definition of an *instantaneous 3-space* is likewise *conventional* and corresponds to a particular (*space-like*) slicing of M^4 made by, and depending on, a particular *fictional observer*;[46] (ii) The light-cone structure leads to a relation between C_{ph} and *physical time* which is *stronger* than the classical relation time-causality. Actually, while the temporal *order*[47] of two C_{ph}-*connectible* events[48] is universally *valid for all observers*, the same *order* of two C_{ph}-*disconnected* events[49] is *observer dependent*; (iii) The meaning (viz. the *attributes*) of any point-event E, in a given world-line of, say, a mass-point *m* is, e.g. that "there is a mass point over there at time *t* for a given *fictional* observer O". Therefore, adding a *new* and *flowing* (viz. *tensed*) time in M^4, in order to describe an alleged 'motion in time' of the mass-point, is *senseless*. Time is *spatialized* and—so to speak—'already there'! (iv) No *transient property* exists or is representable in M^4, no *nowness*, and no *real demarcation* between past and future events. No *potentiality* whatsoever can be (*tensely*) *actualized at a time* in the M^4 tenseless structure, in particular no actual *measurement*, since *all* point-events are 'already' *actualized*[50] and there is no *threshold of actualization*. Nothing is *undecided*. Finally, no *factual* gathering is possible of the instantaneous data which are necessary for a formulation of the *initial value problem* for a system of differential equations. In conclusion, (v) No *living observer* can enforce the Postulate of Freedom within M^4.

Now the question is: how should we interpret this situation *philosophically*? A *literal* interpretation of the RO of STR has been boldly endorsed by Hilary Putnam,[51] who writes (*italics* mine) "I conclude that the problem of the reality and the determinateness of future events is now *solved*. Moreover, it is solved by *physics* and not

---

45 A detailed picture of M^4 can be found in Fig. I.
46 I use the expression *"fictional* observer" since it concerns a mathematical *observer* in the sense of the RO, and not a *living real observer*, belonging to the OW and performing *real physical measures* as actual actions.
47 With respect to the relations *before-than* and *later-than*.
48 That is, lying *one* within the (forward, backward) LC of the *other*.
49 That is, *one* lying outside the LC of the *other*.
50 According to Abner Shimony, they are *"loci of permanent actuality"*; see Shimony 1993a, 141.
51 See Putnam 1967.

by *philosophy*. We have learned that *we live* in a *four-dimensional* and not a three-dimensional world. [...] Indeed, I do not believe that *there are any longer any philosophical* problems about Time; there is only the physical problem of determining the exact physical geometry of the four-dimensional continuum that *we inhabit*".

The *aporetic* character of this interpretation is crystal clear. From all of the above, it follows that M^4 is mainly an ingenious *effective theoretical instrument* for dealing with spatiotemporal physical problems and organizing *spatiotemporal measures*, rather than a representation of the real *Raum* and *Zeit* in which we live and act[52]. Thus we see, in this particular case, that a deep and *non-aporetic* comprehension of the meaning of the physical description of the world requires a reasonable *pragmatic balance and match* between the RO and the OW. In the M^4 case we need—as it were—a living *External Super-Observer* who handles and organizes actual spatiotemporal measures and theoretical inferences by pragmatically exploiting the spatiotemporal knowledge hoarded in the Minkowski picture.[53] This will be actually done in Section 7.1 and represented in Fig. I.

7 From an oversight of Kant's in the Third Antinomy of Reason to the Real Temporality of the Open World

7.1 The oversight

Let me recall some points of Kant's argumentation.[54] In the Notes on Third Antinomy[55] we find: "By freedom in its cosmological meaning, I understand

---

52 Paraphrasing Roberto Torretti, we should say that M^4 is the result of a splendid "creative understanding". Also, Poincaré always considered Minkowski spacetime as a mathematical model and not "un modèle véritablement physique".
53 A concrete example of a *pragmatic* laboratory treatment of relativistic measurements in the case of the *actual* motion of a mass-point is given in Lusanna, and Pauri 2007.
54 The Third Antinomy of Pure Reason: Thesis: "*Causality in accordance with laws of nature* is not the only causality from which the appearances of the world can one and all be derived. To explain these appearances it is *necessary* to assume that there is also *another causality*, that of *freedom*". Antithesis: "*There is no freedom*; every-thing in the world takes place solely *in accordance with laws of nature*"; Kant 1787.
55 See Kant 1787, 466, italics mine.

the power of beginning a state spontaneously (a fact that, even if proved, *is not understood...*). The absolutely first beginning of which we are here speaking *is not a beginning in time,* but *in causality.* If, for instance, I at this moment *arise from my chair, in complete freedom,* without being necessarily determined thereto by the influence of *natural causes,* a new series has its absolute beginning in this event, although *as regards time* this event *is only the continuation of a preceding series* [...] For this resolution and act of mine do not form part of the succession of purely natural effects, and are not a mere continuation of them. In respect of its happening, natural causes exercise over it no determining influence whatsoever. It does indeed *follow* upon them, but *without arising* out of them; and accordingly, *in respect of causality, though not of time,* must be entitled *an absolutely first beginning*". Furthermore: "The principle that all events in the sensible world *stand in thorough-going connection* in accordance with unchangeable laws of nature, which is an established principle of the Transcendental Analytic, *allows of no exception,* [...] for it is only by virtue of it that phenomena can constitute a *nature* and yield objects of experience".

Incredibly, Kant here fails to realize and fully explain the consequences of his premises. For even if the free action had to be referred—according to Kant—to the level of *noumena* (which lies outside the time of appearances), such action in general will have many *phenomenal* consequences, i.e. it will modify the *appearances* in a way that is *totally unpredictable* before the free action. Taking Kant's very example, by arising from the chair he would consequently modify his position in space, the atmospheric flux around his body, etc. By means of a free action, a contemporary Kant could, e.g., detonate a hydrogen bomb, or 50,000 hydrogen bombs and annihilate all life on earth, or even deviate the orbit of an asteroid, etc. Admittedly, such effects are FAPP *local,* but in principle they influence the *whole network of physical causal chains* having their origins before the time of the free action (in a Newtonian sense), or within the relevant forward light cones (in the relativistic sense, see Fig. I).

We see, furthermore, that—from the viewpoint maintained in this Essay—it is *not even true* that the effect of the free action does not have a *new beginning in time* since *nowness* is an essential *meta-physical* issue proper of OW. Then, since the free action takes place in the 'now' of the acting subject, the happening in the framework of appearances is also an *absolutely new* fact of the world that is *unpredictable* as seen in the physical temporal order (see Fig. I). Therefore, it has an *absolute beginning* even in physical time: a new *emergent physical causal chain* has appeared. Of course, while this *tensed experience* in OW is *superimposed* to a certain time t of the underlying B-series structure of the physical description, any attempt to establish *objective* correlations between the 'now' of the free agent and a *quantitative* determination expressed in terms of instants

t of ordinary physical time is prohibited by the RO of STR.[56] Indeed, something *absolutely new* happens for all the actions generated by C_{free}, which take place FAPP continuously in the life of *living organisms*. It is thus clear, that, even if restricted by a multiplicity of limiting factors, the reality of *free actions* radically changes the Kantian *concept* of *law of nature* because it alters the *extension* of its enforceability and, therefore, its *universality*. Indeed, no system of lawful connections between events in OW can *presently* be *thorough-going*, contrary to Kant's dictum in his Observation on the Antithesis.[57] The remaining universality of physical laws is no longer representable as a closed totality of physical causal connections and turns out to be decomposed into *sectional pieces*, between the spatiotemporal *origins* of *emergent* chains, despite coherently.[58]

I shall attempt an expressive kind of relativistic picture of the consequences of Kant's oversight. To this effect, adopting the *pragmatic match* considered above, I shall exploit—so to speak—a *projection* from OW into the RO of STR. This is done by an External Super-Observer (ESO) in Fig. I in a way which I am going to expound. For the sake of clarity, I shall adopt a special terminology as follows. A C_{free}-action of an agent F generates (obviously in his proper *now*) a new *state of affairs* around a certain event A of M^4 [coordinatized, say, as A=(ct,**x**) by an inertial *fictional* Lorentz observer K] and can be *pragmatically* represented as follows. The event A becomes the C_{ph}-*acausal* origin of an emergent C_{ph}-causal chain (or a set of emergent C_{ph}-*causal chains*) within the *forward* light-cone Γ_A of A. Such *emergent* C_{ph}-*causal* chains modify *all* the world-lines corresponding to C_{ph}-*causal* chains with C_{ph}-*causal* origin in an event, say B (C_{ph}-*causally disconnected* from A), within the intersection $\Gamma_A \cap \Gamma_B$. The new state of affairs in the intersection can also be viewed as a *local* modification of *initial* conditions of the C_{ph}-*causal* chains with origin in B, *actualized* at an event E of $\Gamma_A \cap \Gamma_B$ by the C_{free}-action of the agent F in A. This shows with crystal clear evidence the aporetic relationships between the RO of STR, as *literally interpreted*, and the reality of free-will in OW. The *actualization* in E requires a *pragmatic* interpretation of M^4.

---

56 A relevant exception, however, is provided by a possible correlation with the (physical) cosmic time order (see later).
57 Note that I am just *reversing* the current exploitation of Kant's Third Antinomy *against* the mind-body dualism.
58 Note that this holds in addition to the fact that *no system of lawful connections* among given actual events can *presently* be *thorough-going* in OW, independently of C_{free}. As Torretti remarks: "Any definite set of events must sport *vacant slots* for connections 'according to unchangeable natural laws' with still other events"; see Torretti 1999.

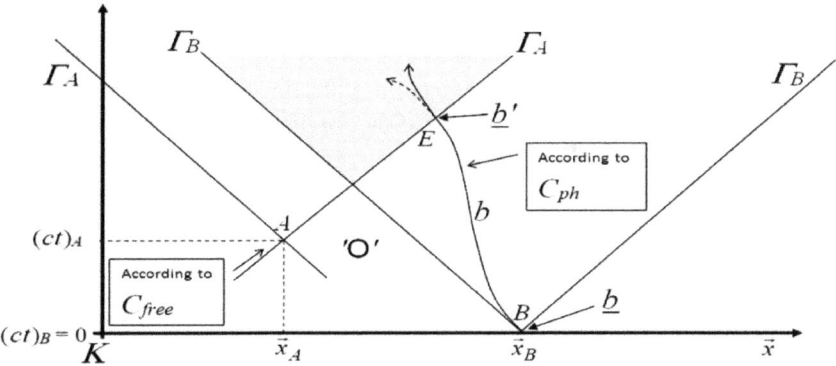

FIG.I: UPDATED MINKOWSKI SPACE-TIME
An External Super-Observer's [ESO] view
of the relativistic C_{ph}-causal consequences of a C_{free}-causal action

Legend:

'O' = External Super-Observer:

Γ_A = forward light-cone of event A, where a free action by 'O' is operated according to C_{free} (obviously on the now" of the agent), at the physical time t_A for the Lorentz fictional observer K.

Γ_B = forward light-cone of event B at the physical time t_B=0 of K. b is a generic physical causal chain generated at B, with initial conditions $\underline{b}$. Note that the events A and B are C_{ph}-causally disconnected.

The intersection $\Gamma_A \cap \Gamma_B$ of the forward light-cones is the spatiotemporal region where the phenomenal physical effects of the free action at A influence the physical causal chains originated at B. $\underline{b}'$ are the new initial conditions induced by the free action at the event E on the pre-existing physical causal chain b, which turns into b' within the intersection.

Clearly, the new initial conditions $\underline{b}'$ for b at E are completely unpredictable in terms of $\underline{b}$ (this is precisely Kant's oversight). Yet, after the free-agent choice at A, and until after another free-action, the world is locally describable in terms of the laws of RO without any threat to the physical description.

7.2 The Real Tran-subjective Temporality of the Open World

We see in Fig. I that all *physical effects* of the *free action* at the event A upon a C_{ph}-*causal chain* originated at the event B (C_{ph}-*causally disconnected* from A), and checked at *the event E*, are completely *unpredictable* from the 'view-point' of B. Now, in order to obtain a relativistic 'super- representation' including the effects of *free actions*, the ESO must *update* M^4 *tensely* for *every free action*, while *leaving* the updated copy *tenselessly identical* until the subsequent updating. In this way, C_{free} projects an *additional temporal structure* on spacetime, constituted by the *tense* inclusion of *updated copies* of M^4 corresponding to *each free action* (this structure will indeed be continuous if the set of free actions is such).

As already anticipated, I call the *tensed transubjective* time of the continuously-updated spacetime *Real Time* of the *Open World*, or *Time of Becoming*. The C_{ph}-*acausal* origins of the *emergent* C_{ph}-*causal* chains appear in the experienced real world as *Traces* of *Becoming*. The intrinsic, *tensed*, directionality of phenomenological time overlaps the *tenseless* anisotropy of physical time which, as well-known, is only a *de facto* but not a *de lege* fundamental feature.

On the other hand, we, as *updating* ESOs, are the same agents that live—so to speak—*outside* M^4, prepare physical states, perform measures, elaborate physical theories and are obliged to use a *tensed* discourse. Thus, *Real Time* is inherent in the emerging structure of OW, while the physical description of the world in terms of the RO provides a *model* of the *world* in which the temporal structure can be shown in a *tenseless* form only. Certainly, due to relativistic reasons, the *real time order* is essentially *local*, though *objectively* localized around any of the *a-causal* origins of *emergent* C_{ph}-*causal* chains. However, by exploiting, *e.g.* the physical *cosmic time* of the co-moving Robertson-Walker global inertial frame, the *emergent* C_{ph}-*causal chains* could be arranged in a global *anisotropic* time-like *cosmic order*. Admittedly, cosmic time has a mere descriptive privilege. Yet, this arrangement appears to be a physical cognate of Whitehead's "cosmic unison" (1955).

In conclusion, *Becoming* is nothing else than the *actualization of potentialities*. It does correspond to a *change of time*, and not to a *change in time* so that it is necessarily only latent in the *tenseless* physical description. As Richard Gale (1960) put it, "What time enables us to say is exactly what cannot be said about time". Finally, living organisms appear to be the true *disclosers* of the *traces of becoming* as *Real Time* of the *Open World*.[59]

8 Concluding remarks

I started by saying that I had no intention of proposing any *positive* metaphysics. Although my view of Free Will would entail a form of *implicit psycho-physical dualism*,[60] I will not comment on it further. Yet, a big ontological *genetic problem* surely lurks in the background here. The only frame in which I can grasp *ontological emergence* is through the (cosmological) reality of *potentiality*. Concerning *potentiality*, I wish to recall the original works on *emergence* by logical-positivists

---

[59] We could even say that their singularity is that of owning *special temporal properties* as compared to what we call inanimate matter.
[60] Even without dualism of *substances*. See, e.g., Corradini 2008, 2011.

and physicalists such as Feigl (1958), Meehl and Sellars (1956). In this connection, they introduced a remarkable distinction concerning physical language. Precisely: "1) *Physical-1*: an event, entity or process is *physical-1* if it belongs in the coherent and adequate descriptive and explanatory account of *our* spatio-temporal *causal world*. 2) *Physical-2*: an event, entity or process is *physical-2* if it is definable in terms of theoretical primitives adequate to describe completely the actual states, *though not necessarily the potentialities*, of the universe *before the appearance of life*". These authors felt obliged, as it were, to introduce *potentialities* in order to deal with *emergence*, by adding in the sequel that "*Physics-2* should possess the capacity to account for *objective phenomenal properties* which *now*, after the appearance of *living organisms* and *subjects* could be only *experienced* through *introspection*". This is purely metaphoric thinking, of course, yet discloser of some deep ontological necessity of characterizing the issue of emergence of subjectivity by recourse to potentiality. The core point is that physics deals with physical systems, objects and processes and their RO, which are very sophisticated concepts, but never *directly* with 'matter', which has always been and still remains a philosophical concept. As Bertrand Russell stressed, 'matter' is *inscrutable*. Of course, we here have a huge *explanatory gap*. An *explanatory* gap of some sort, however, seems to be present any time a bridge is laid from the *scientific image* to *subjectivity* in any of its forms. In this connection, I cannot here overlook Leibniz's definition of 'matter': "Mens *momentanea*, atque *sive recordatione*". Surely, he was not a reductionist.

APPENDIX A: The aporetic nature of reductionist physicalism, a case study: Weinberg's *Deterministic Grand Reductionism*

"We should distinguish between reductionism as a *program for scientific research* and what I call 'Grand Reductionism', which is an (ontological) *view of nature* [...] Phenomena like *mind and life do emerge*. The rules they obey are not independent truths, but follow from scientific principles at a deeper level. Apart from initial conditions and historical *accidents* that by definition cannot be explained [this indeed must be a leak in the argument!!!], the *nervous systems of George and his friends have evolved* to what they are *entirely* because of the *principles of macroscopic physics and chemistry*, which in turn are what they are *entirely* because of *principles of standard model* of elementary particles [...] *Determinism* is logically distinct from *grand reductionism*, but the two doctrines tend to go together because the *reductionist goal of explanation* is tied in with the *determinist idea of prediction*; we test our explanations by their power to make successful predictions" (Weinberg 1995).

Weinberg and his "imaginary friend" George

I will use Jaegwon Kim's (1998) terminology: *mental properties* are physically realized as "second order properties" causally generated out of "first order" *non-mental* properties.

FIG.II: A REALISTIC HAPPENING INVOLVING STEVEN WEINBERG AND GEORGE

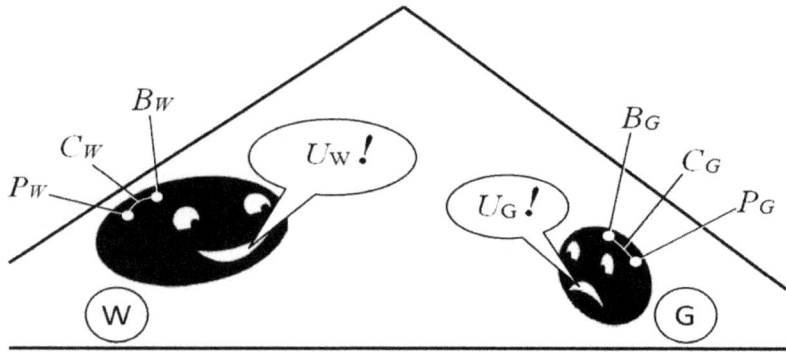

Legend:

B_W = physical state of Weinberg's brain; B_G = physical state of George's brain;
P_W = Weinberg's brain's "second order properties"; P_G = George's brain's "second order properties";
C_W = causal specification of P_W; C_G = causal specification of P_G;
U_W = Weinberg's utterance: "**GRAND REDUCTIONISM IS TRUE!**"
U_G = George's utterance: "**GRAND REDUCTIONISM IS BULLSHIT!**"

→ Weinberg believes that: C_W C_{ph}-causally determines P_W in such a way that U_W expresses a real truth.
→ Weinberg must also believe that: C_G C_{ph}-causally determines P_G in such a way that U_G *expresses a real falsity*.

Note, on the other hand, that George *is not self-contradictory* since he can rely in principle upon a notion of truth bypassing DGR. Let me go on: In fact, I guess that Weinberg, having heard George's utterance, would add to his first utterance the following specification:
"The causal generation of a belief does not, of itself, detract in the least from its truth".[61]
However, he is *de facto* also accepting the validity of the specular sentence:
"the causal generation of a belief does not, of itself, detract in the least from its falseness".

---

[61] This sentence has been indeed presented to my attention by Adolf Grünbaum in a private discussion. At the time I had no occasion to reply with the *specular* sentence that I am now writing here. For Grünbaum's compatibilist view see Grünbaum 1972.

Yet the point is that, since in DGR (intended as an *ontological view of nature*) there is no other mechanism at work than the physical causation (C_{ph}) of facts, beliefs and whatever else, there is no *sufficient reason* for a belief or a sentence to be *true* or *false*. Even more—within Weinberg's ontology—no *conceptual room* exists, or can be added, for *truth* and *falseness in general*, no sensible notion of *meaning* and *truth* validating Weinberg's utterance in front of other, *ontologically equivalent*, utterances, i.e. there is in particular no *rational discrimination* between Weinberg's and his opponent's views. In conclusion: as soon as Weinberg *states*—against George—that DGR is *the only ontologically admissible view of nature*, he falls into OPA: his ontology cannot justify the *phenomenology of meaning* he *is using* against his opponent's view.[62]

APPENDIX B: Language in the *"event"- framework"*

The so-called *A* and *B series*[63] of temporal determinations are: *A*: characterized by *monadic,* so-called *tensed,* terms: "The event E is present, past, future". Truth-values of A-sentences subtend a *subjective temporal perspective* of the utterer ("Today, in Milan, it is raining"). In general A-sentences about future contingents do not satisfy the Bivalence Principle, i.e. they can be neither false nor true (recall the sentences on future happenings in Aristotle's battle of Salamina). *B*: characterized by *bi-adic relational,* so-called *tenseless,* terms: "E_1 is *before, after, simultaneous with* another event E_2". Truth-values of B-sentences are *independent of any temporal perspective* ("On June 4th 2013, it is raining in Milan"). This is true or false for any time. (Bivalence Principle always satisfied.) Real events are ordered by B-series relations with the set (call it *W*) of 'all real events in the world': Plato does really exists *tenselessly* in the sense that—unlike the *hyppogriph*— he belongs to *W*. Note that from a logical point of view, i.e. with reference to what can be captured and represented in language, *tensed* and *tenseless* sentences are *identical* (if the former is uttered at the right time); yet *tensed* sentences contain additional information of *existential* nature which is essential for an *action* to be performed *now* (e.g. opening an umbrella if *now* it is raining).

---

62 More on this analysis can be found in Pauri 1997a, 1997b, and 2008, 168–169.
63 See McTaggart 1908.

Acknowledgments

I wish to warmly thank my dear friend Prof. Mario Casartelli for a critical and illuminating reading of the manuscript.

References

Bohm, D. 1980, *Wholeness and the Implicate Order*, London: AIR Paperbacks.
Carnap, R. 1928, *Die logische Aufbau der Welt: Scheinprobleme in der Philosophie*. New English edition by R.A. George 2003, *The Logical Structure of the World and Pseudoproblems in Philosophy*, Chicago, IL: Open Court Classics.
Conway, J.H., and Kochen, S. 2006, "The Free-Will Theorem", *Foundations of Physics*, 36 (10): 1441–1473.
Corradini, A. 2008, "Emergent Dualism", in A. Antonietti, A. Corradini, and J. Lowe (eds.), *Psycho-Physical Dualism Today. An Interdisciplinary Approach*, Lanham, UK: Lexington Books, 185–209.
Corradini, A. 2011, "Philosophy and Neuroscience", in C. Kanzian, W. Löffler, and J. Quitterer (eds.), *The Ways Things Are. Studies in Ontology*, Frankfurt a.M.: Ontos Verlag, 203–219.
Dennett, D. 1979, "The Absence of Phenomenology", in D. Gustafson and B. Tapscott (eds.), *Body, Mind, and Method*, Dordrecht: Kluwer.
Dennett, D. 2003, *Freedom Evolves*, New York: Viking Books.
Feigl, H. 1958, "The 'Mental' and the Physical'", in H. Feigl, M. Scriven, and G. Maxwell (eds.) 1967, *Minnesota Studies in the Philosophy of Science*, Vol. 2, *Concepts, Theories and the Mind-Body Problem*, Minneapolis: University of Minnesota Press.
Gale, R. 1968, *The Language of Time*, London: Routledge and Kegan Paul.
Grünbaum, A. 1972, "Free-Will and Laws of Human Behaviour", in H. Feigl, W. Sellars, and K. Lehrer (eds.), *New Readings in Philosophical Analysis*, New York: Appleton-Century-Crofts, 605–627.
Hawking, S.W. and Ellis, G.F.R. 1973, *The Large Scale Structure of Space-Time*, Cambridge, UK: Cambridge University Press.
Heisenberg, M. 2009, "Is Free Will an Illusion?", *Nature* 459: 164-165.
Kane, R. 1996, *The Significance of Free-Will*, Oxford: Oxford University Press.
Kant, I. 1785, *Grundlegung zur Metaphysik der Sitten*. English translation by T. Kingsmill Abbott, edited with revisions by L. Denis 2005, *Groundwork for the metaphysics of morals*, New York: Broadview Press.
Kant, I. 1787, *Kritik der reinen Vernunft*, 2nd edition, Riga: Hartknoch, 1st edition 1781. English translation by N. Kemp Smith 1963, *The Critique of Pure Reason*, London: Macmillan.
Kim, J. 1998, *Mind in a Physical World. An Essay on the Mind-Body Problem and Mental Causation*, Cambridge, Mass.: The MIT Press.
Laplace, P.S. 1812, *Théorie Analytique des Probabilités*, Paris: Courcier.
Longo, G. and Montévil, M. 2014, *Perspectives on Organisms, Lecture Notes in Morphogenesis*, Berlin: Springer.

Lusanna,L., and Pauri, M. 2007, "Dynamical Emergence of Instantaneous 3-Spaces in a Class of Models of General Relativity", in V. Petkov (ed.), *Relativity and Dimensionality of the World*, Berlin: Springer.

McTaggart, J.E. 1908, "The Unreality of Time", *Mind* 17 (4): 457–473.

Meehl, E. & Sellars, W. 1956, "The Concept of Emergence", in H.Feigl and M.Scriven (eds.), *Minnesota Studies in the Philosophy of Science*, Vol. 1, *The Foundations of Science and the Concepts of Psychology and Psychoanalysis*, Minneapolis: University of Minnesota Press.

Meixner, U. 2004, *The Two Sides of Being: a reassessment of psycho-physical dualism*, Paderborn: Mentis.

Meixner,U. 2006: "Consciousness and Freedom", in A. Corradini, S. Galvan, and E.J. Lowe (eds.), *Analytic Philosophy Without Naturalism*, London: Routledge, 183–196.

Meixner, U. 2008, "New Perspectives for a Dualistic Conception of Mental Causation", *Journal of Consciousness Studies* 15 (1): 17–38.

Nagel, E. 1939, *Principles of the Theory of Probability*, Oxford: Oxford University Press.

Nagel, T. 1987, *What Does it All Mean ? A Very Short Introduction to Philosophy*, Oxford: Oxford University Press.

Pauri, M. 1991, "The Universe as a Scientific Object", in E. Agazzi and A. Cordero (eds.), *Philosophy and the Origin and Evolution of the Universe*, Synthese Library, Vol. 217, Dordrecht, Holland: Kluwer Academic Publishers, 291-339.

Pauri, M. 1997a, "La Descrizione Fisica del Mondo e la Questione del Divenire Temporale", in G. Boniolo (ed.), *Filosofia della Fisica*, Milano: Bruno Mondadori, 245–333.

Pauri, M. 1997b, "Time: The Physical Worldview and Becoming", in J. Faye, U. Scheffler, and M. Urchs (eds.), *Perspectives on Time*, Boston Studies in the Philosophy of Science, n. 189, Dordrecht, Holland: Kluwer Academic Publishers, 267–297.

Pauri, M. 1998, "Do Living Organsims Possess Peculiar Temporal Properties?", in J. Tojo Suarez (ed.), *International Conference on Philosophy of Science: Philosophy of Biology*, Vigo: Universidade de Vigo Press, 163–189.

Pauri, M. 2001, "Perennial Cosmology", Invited Presentation at the Workshop SSQ-II: Physics and Cosmology, UNESCO, Paris, January 9–12.

Pauri, M. 2008, "Time, Physics and Freedom: at the Roots of Contemporary Nihilism", Académie Internationale de Philosophie des Sciences, *Time in the Different Scientific Approaches*, Colloques de Cerisy, 4-9 Octobre 2007; *Epistemologia*, Vol. 14, Special Issue: 155–192.

Pauri, M. 2011, "Epistemic Primacy vs. Ontological Elusiveness of Spatial Extension: Is There an Evolutionary Role for the Quantum?", *Foundations of Physics* 41: 1677–1702.

Popper K.R., and Eccles, J. 1977, *The Self and Its Brain. An Argument for Interactionism,* Berlin: Springer.

Putnam, H. 1967, "Time and Physical Geometry", *The Journal of Philosophy*, 64 (8): 240–247; reprinted in *Mathematics, Matter and Method: Philosophical Papers*, Vol. 1, 1979, Cambridge, UK: Cambridge University Press.

Shimony, A. 1993a, "Events and processes in the quantum world", in *Search for a Naturalistic World View*, Vol. 2, *Natural Science and Metaphysics*, Cambridge, UK: Cambridge University Press, 140–162.

Shimony, A. 1993b, "The Transient *Now*", in *Search for a Naturalistic World View*, Vol. 2, *Natural Science and Metaphysics*, Cambridge, UK: Cambridge University Press, 271–287.

Torretti. R. 1999, *The Philosophy of Physics*, Cambridge, UK: Cambridge University Press.

Torretti, R. 2000, "Spacetime Models for the World", *Studies in History and Philosophy of Modern Physics*, 31, 171–186.

Weinberg, S. 1995: http:/www.idt.mdh.sc/kurser/ct3340/ht02/Reductionism_Redux.pdf.
Whitehead, A.N. 1955: *An Enquiry Concerning the Principles of Natural Knowledge*, Cambridge, UK: Cambridge University Press.
Wigner, E.P. 1960, "The Unreasonable Effectiveness of Mathematics in Natural Sciences", *Commun.Pure Appl. Math.*, 13: 1–14.

Robert Kane
Quantum Physics, Action and Free Will: How Might Free Will be Possible in a Quantum Universe?

1 Rethinking the Problem of Free Will

When I first began thinking about issues of free will nearly fifty years ago, the contours of free will debate were simpler than today. The common assumption was that if you had scientific leanings, you would naturally be a compatibilist about free will (believing it to be compatible with determinism), unless of course you were a skeptic or hard determinist denying free will altogether. Compatibilism was the default position for philosophers and scientists and still is today, though to a lesser degree. And if by contrast you were a libertarian about free will, believing in a free will that was incompatible with determinism, it was assumed back then that you must inevitably appeal to some kind of obscure or mysterious forms of agency or causation to make sense of it—to uncaused causes, immaterial minds, noumenal selves, prime movers unmoved, or other examples of what P. F. Strawson called the "panicky metaphysics" of libertarianism in his influential 1962 essay "Freedom and Resentment".

I began thinking about free will shortly after Strawson's essay appeared, when my philosophical mentor at the time, Wilfred Sellars, challenged me to reconcile a traditional incompatibilist or libertarian free will with modern science. As a compatibilist about free will, Sellars doubted a libertarian free will could be accounted for without reducing it to mere chance or appealing to what Strawson had dubbed panicky metaphysics. Though Sellars granted that free will in some sense was an integral part of what he notably called the *manifest image* of the world, he did not believe a traditional libertarian free will could be reconciled with what he called the *scientific image* of the world; and he challenged me to show otherwise.

Meeting this challenge turned out to be a longer task than I expected back then as a young graduate student—over forty years now and still ongoing. The reason, as I found out, is that it would require rethinking nearly every facet of the traditional problem of free will from the ground up.

Another feature of the situation in the 1960s was the widespread conviction that Quantum Theory "would be no help" in explaining and justifying free will.

This was obvious for *compatibilists*, the majority group among philosophers and scientists. Since compatibilists believed free will was compatible with determinism, they believed the indeterminism implied by standard interpretations of quantum physics was not needed for free will and indeed would not enhance our freedom but diminish it.

More surprising was the fact that many *libertarians* about free will (who believed free will was *incompatible* with determinism) were also inclined to believe that quantum physics was no help in explaining free will. There were two reasons for this widespread conviction. First, there was the widespread conviction that quantum indeterminacies would be damped out in macroscopic systems such as the brains and bodies of living organisms and would play no significant role. But, second, and even more important, doubts were widespread that even if quantum indeterminacies did under certain conditions impact macroscopic behavior, this would be no help with free will. If a choice or decision resulted from undetermined events in an agent's brain, it was widely believed, its occurrence would be a matter of chance or luck, not a free and responsible choice. This conclusion might be avoided by appealing to mysterious agencies outside the natural order, as Kant, for example, had appealed to a noumenal self outside space and time to explain free will. But most scientists and philosophers rejected such appeals to non-natural agencies. And indeterminism itself in the brain or behavior without such additional appeals, they believed, would not enhance freedom and control over events, but would diminish freedom and control.

It was an old story. The Epicurean philosophers of old had said that if the atoms did not sometimes "swerve" in chance ways there would be no room in nature for free will. But their many critics, such as the Stoics, cried out in opposition: How could the chance swerve of atoms help with free will? Free will is not mere chance.

In opposition to this widespread consensus fifty years ago that quantum theory would be "no help" with free will, my view was that quantum theory *was* important for free will and must play some role or other. But the challenge was to make sense of what that role might be without reducing free will either to mystery or to mere chance.

2 Rethinking the Compatibility Question: Free Action and Free Will

Meeting this challenge, as I said, would turn out to require rethinking many facets of the traditional problem of free will. I first want to discuss two aspects of this rethinking that were crucial to the development of my view. The first concerns the notion of *free will* itself and the meaning of the phrase in which the term free will most often occurs in everyday affairs, "acting *of one's own free will*". Understanding this phrase requires reflecting on the relation between freedom of will and freedom of action.

There had been a tendency in the modern era beginning in the seventeenth century and coming to fruition in the twentieth century to reduce the problem of free will to a problem about free action. The expression "free will" is often said to be merely an honorific title we give to the problem of free action with a bow to its historical past. Worthy thinkers like Ludwig Wittgenstein and Gilbert Ryle convinced many philosophers that the *will* and *acts of will* were suspect medieval notions that should go the way of witches and phlogiston. Talk of the will itself suggested an inner homunculus or prime mover unmoved and hence more mystery.

I have resisted this modern trend to reduce the problem of free will to merely a problem of free action because I think it oversimplifies the ancient problem. John Locke had said in the seventeenth century that the issue was about the freedom of the *agent,* not the freedom of the *will,* and this view gained much currency. But I think Locke's older contemporary Bishop Bramhall, who had debated the issue with Thomas Hobbes, was closer to the truth when he said that the freedom of the agent was *from* the freedom of the will. There is no contradiction in saying that the problem is both about free agency, as Locke said, and free will, since the freedom of the will is an important aspect of free agency and, crucially, it is that particular aspect of free agency which raises very deep philosophical problems.

In the last fifty years, the only other philosopher known to me who has emphasized the importance of distinguishing freedom of will from mere freedom of action is Harry Frankfurt, whose writings on the subject have been very influential (see, especially, 1971). But we have diametrically opposed views about the meaning of free will and about of crucial phrase "acting of one's own free will". Frankfurt thinks we act "of our own free will" when we act from a will with which we *identify* and to which we are *wholeheartedly* committed without any *ambivalence.* In short, we have free will when we have the will or first-order desires we want to have and are wholeheartedly committed to them. For me, by contrast, to act of one's own free will is to "act from a will that is to some significant degree *a will of one's own free making*". My view is historical (emphasizing how we *came to have* the will we do have); Frankfurt's is not historical. His is view is compatibilist, mine is not.

Many years ago, I wrote a short letter to Frankfurt which mentioned several criticisms of his view. The known criticism was that whether or not we were wholehearted about the will or first-order desires we have, rather than ambivalent about them, could on his view be entirely a matter of social conditioning over which we had no control. The second criticism was this: If freedom of will is being wholehearted in your commitments to what you presently will or desire without any ambivalence, then no one could ever get from ambivalence to wholeheartedness "of their own free will". For they wouldn't *have* free will on his view until they got there.

His response began by saying "while many people think I'm crazy on this point", I believe it doesn't matter how you came to have free will or wholeheartedness. It may only be because of luck or good fortune or upbringing or accidental factors or even social conditioning. And no, he added, you can't get from ambivalence to wholeheartedness of your own free will in any deeper sense of that term. All that matters is that you have free will or wholeheartedness no matter how you got it, for it is a good which makes life go well. This forthright answer did clearly frame the debate between us about what free will might involve.

What then of my own incompatibilist or libertarian account of the expression "of one's own free will"? I believe it means what was traditionally meant by it, namely a "will of *one's own* free making", rather than a will wholly formed by factors other than oneself and over which one did not have control. And this interpretation brings us to another way in which I have argued over the past several decades that the free will issue required rethinking.

3 Alternative Possibilities and Ultimate Responsibility

The preceding remarks concern the role of the term "*will*" in the expression "acting of one's own free will". We must now turn to the meaning of the term "*free*" in that expression. The meaning of "free" in acting "of one's own free will" that is usually thought to cause a conflict with determinism is normally understood to be the requirement that the free agent "*could have done otherwise*" or had *alternative possibilities*, a requirement I call AP. Most recent and historical arguments for the incompatibility of free will and determinism have appealed to this requirement AP in one way or another and have revolved around questions of whether an agent's power to do otherwise is or is not compatible with determinism. Contentious debates about these arguments for incompatibilism from AP in modern times, as is well known, have frequently tended to stalemate over differing interpretations of what it means to say that agents have the *power* or *ability* to act otherwise or "*could* have done otherwise".

I have argued that persistent disagreements over the meanings of these terms are symptoms of a deeper problem—namely that focusing on the requirement of alternative possibilities or AP alone is *too thin* a basis on which to rest the case for the incompatibility of free will and determinism. This does not mean that alternative possibilities have no role to play in debates about free will and determinism, as some compatibilists would have us believe. But it does mean that these debates cannot be resolved by focusing on alternative possibilities *alone*.

Fortunately there is another place to look for reasons that free will might conflict with determinism. In the long history of free will debate, I have argued that one can find another condition fueling incompatibilist intuitions even more important than AP or alternative possibilites. I called it the condition of "ultimate responsibility" or UR. The basic idea is this: to be ultimately responsible for an action, an agent must be responsible for anything that is a sufficient reason (condition, cause or motive) for the actions occurring. If for example a choice issues from and can be sufficiently explained by, an agent's character, motives and intentions, i.e. the agent's pre-existing will (together with background conditions), then to be ultimately responsible for the choice, the agent must be at least in part responsible by virtue of choices or actions voluntarily performed in the past for having the character, motives and purposes he or she now has. Compare Aristotle's claim (1915) that if a man is responsible for wicked acts that flow from his character, he must at some time in the past have been responsible for forming the wicked character from which these acts flow.

Note that this condition UR does not require that we could have done otherwise, AP, for *every* act done "of our own free will," thus vindicating various philosophers, Frankfurt included, who insist that we can be held morally responsible for many acts even when we could not have done otherwise. But that's only half the story. For UR does require that we could have done otherwise with respect to *some* acts in a past life histories by which we *formed* our present characters or wills. I call these self-forming actions or SFAs.

Often we act from will already formed, but it is "our own free will" by virtue of the fact that we formed the will from which we act by other choices or actions in the past (SFAs) for which we could have done otherwise. If this were not so, I argue, *there is nothing we could have ever done differently in our entire lifetimes to make ourselves different than we are*—a consequence, I believe, that is incompatible with our being (at least to some degree) ultimately responsible for what we are.

This is one reason why I have argued that the tendency in the modern era from Hobbes and Locke onward to reduce the problem of freedom of the will to just a problem of free action is a mistake and oversimplifies the problem. Free will is not just about free action. It is about *self-formation*, about the formation of our *wills* (our characters, motives and purposes), or how we got to be the kinds of persons we are, with the wills we now have. Were *we* ultimately responsible to some degree for having the wills we do have, or can the sources of our wills be completely traced backwards to something over which we had no control such as fate or the decrees of God, or heredity or environment, upbringing or social conditioning or hidden controllers, and so on? Therein, I believe, lies the core of the traditional problem of free will.

Finally, if the case for the incompatibility of free will and determinism cannot be made on AP alone, it can be made if UR is added. If agents must be responsible to some degree for anything that is a sufficient cause or motive for their actions, then an impossible infinite regress of past actions would be required unless some actions in an agent's life history—self-forming actions or SFAs—did not have either sufficient causes or motives and hence were undetermined.

4 The Intelligibility Question

But this different approach to the incompatibility question through UR raises a host of further questions about free will, including how actions or choices lacking both sufficient causes and motives could themselves be free and responsible actions, and how if at all such actions could exist in the natural order where we humans exercise our freedom.

These are versions of what I have called the Intelligibility Question about incompatibilist free will. This question is connected, as noted earlier, to an ancient dilemma: If free will is not compatible with determinism, it does not seem to be compatible with indeterminism either. Indeterminism means same past, different possible futures. But how is it possible, one might ask, that different actions could arise voluntarily and intentionally from the same past without occurring merely by luck or chance? If a choice occurred as a result of a quantum jump or other undetermined event in one's brain, would that amount to a free and responsible choice? From such thoughts and others flow many traditional arguments to the effect that undetermined choices or actions would be "arbitrary", "capricious", "random", "irrational", "uncontrolled", "inexplicable", or "mere matters of chance or luck", and hence not free and responsible actions at all.

The usual answers of libertarians to such objections in the past, as noted, have involved appeals to various extra factors, from noumenal selves and immaterial minds, to nonevent agent causes etc. (Strawson's "panicky metaphysics"). It was assumed that some such extra factors or other were needed to fill the causal gaps in nature that mere indeterminism left unfilled. I long ago became disenchanted with all such appeals and thought one had to try something new. But where to go if one is to avoid such appeals?

First, let us be clear that there has to *be* some genuine indeterminacy in nature in appropriate places to influence human behavior, if any view of free will requiring indeterminism is to make sense. And whether there is such is an empirical and scientific question that cannot be settled by a priori or philosophical reasoning. The question we are addressing here, however, is a philosophical one that has boggled people's minds for centuries, from the time of the Epicureans onward: What could one *do* with indeterminism, assuming it was there in nature in the right places, in the brain for example, to make sense of free will as something other than *mere* chance or randomness and without appealing to mystery?

This is the question I will be addressing. But first a word about the empirical and scientific issues. When I began dealing with free will in the 1960s it was almost universally assumed that the brain could be viewed as a deterministic mechanism, albeit a very complex one, in which quantum indeterminacies involving elementary particles played no significant role. This assumption is still undoubtedly the majority one. But in the past two decades there has been considerably more openness and discussion about the possibility that indeterminism might play a significant role in human and animal behavior and in the functioning of the brain. Many of the contributions to this volume provide evidence of the new found openness to this possibility.[1] And other recent writings by biologists and neuroscientists provide further evidence.[2]

As Mark Balaguer has argued in his 2010 book, *Free Will as an Open Scientific Problem*, the claim that current "neuroscience treats all neural processes as deterministic is straightforwardly false. Current neuroscientific theory treats a number of different neural processes probabilistically, and any decent textbook on neuroscience will point this out. For instance, synaptic transmission and spike firing [the firings of individual neurons] are both treated probabilistically" (pp. 163–164). One textbook (Dayan, and Abbott 2001) puts these points as follows:

> Synaptic] transmitter release is a stochastic process. Release of transmitter at a presynaptic terminal does not necessarily occur every time an action potential arrives and, conversely, spontaneous release can occur even in the absence of [the arrival of] an action potential. (p. 179) [...] Because the sequence of action potentials generated by a given stimulus varies from trial to trial, neuronal responses are typically treated statistically or probabilistically" (p. 9).

"It is worth noting", Balaguer adds, "that some aspects of the indeterminacies of both of these processes [synaptic transmission and spike firing] are caused by the indeterminacy inherent in another process, namely, the opening and closing of ion channels, which are essentially gates that let charged ions in and out of cells.

---

1 For discussion of some of the physics and neuroscience issues involved, see, e.g., papers in this volume by Stapp, Pauri, Hameroff, Mohrhoff, Jedlicka.
2 A small sampling of the relevant literature would include Beck, and Eccles 1992; Penrose 1994; Glimcher 2005; Atmanspacher, and Rotter 2008; Brembs 2010; Heisenberg 2013.

Now, to be sure, by treating these processes probabilistically, neuroscientists do not commit themselves to the thesis that, in the end, they are genuinely indeterministic. But the important point here is that they aren't committed to determinism either. The question of whether these processes are genuinely indeterministic simply isn't answered by neuroscientific theory".

Balaguer goes on to quote several neuroscientists, including Dayan, one of the authors of the text just cited, who says that "people would argue that there are good thermal reasons to think that [the opening and closing of ion channels] is truly random. Thus, short of philosophical debates about hidden variables for all forms of randomness in physics, this is some fundamental randomness for which people nearly have evidence". And Sebastian Seung, a neuroscientist at MIT, says that "The question of whether [synaptic transmission and spike firing] are 'truly random' processes in the brain isn't really a neuroscience question. It's more of a physics question, having to do with statistical mechanics and quantum mechanics".[3]

Now none of this proves that indeterminism plays a significant role in the brain. We are a long way from demonstrating that. But neither does it show that the brain operates on only or strictly deterministic laws. It is enough for our present purposes to say, as Balaguer does, that it is an "open scientific problem".[4] As it happens, on the libertarian account of free will to be presented, one does not need large-scale indeterminism in the brain, in the form, say, of macro-level wave function collapses. Minute indeterminacies in the timings of firings of individual neurons would suffice, because the indeterminism in my view plays only an interfering role, in the form of background noise, as we shall see. In other words, indeterminism does not have to "do the deed" on its own, so to speak. One does not need a downpour of indeterminism in the brain, or a thunderclap, to get free will. Just a sprinkle will do.

5 Steps Toward a Solution

With that said, let us return to the Intelligibility Question as previously stated: What could one *do* with indeterminism, assuming it was there in nature in the right places, in the brain for example, to make sense of free will as something other than mere chance or randomness and without appealing to mystery?

---

[3] Quoted in Balaguer 2010, 34 from correspondence with Seung.
[4] On the matter of its being an "open scientific problem", see also Roskies (in press).

The first step in answering this question is to note that indeterminism need not be involved in all acts done "of our own free wills", as argued earlier. Not all of them have to be undetermined, but only those choices or acts in our lifetimes by which we make ourselves into the kinds of persons we are, namely, the "self-forming" actions (SFAs) of earlier sections.

Now I believe such self-forming actions (SFAs) occur at those difficult times of life when we are torn between competing visions of what we should do or become; and they are more frequent in everyday life than one may think. Perhaps we are torn between doing the moral thing or acting from ambition, or between powerful present desires and long term goals, or we are faced with difficult tasks for which we have aversions. In all such cases, we are faced with competing motivations and have to make an effort to overcome temptation to do something else we also strongly want. At such times, there is tension and uncertainty in our minds about what to do, I theorize, that is reflected in appropriate regions of our brains by movement away from thermodynamic equilibrium—in short, a kind of stirring up of chaos in the brain that makes it sensitive to micro-indeterminacies at the neuronal level.[5] The uncertainty and inner tension we feel at such soul-searching moments of self-formation would thereby be reflected in the indeterminacy of our neural processes themselves. What is experienced personally as uncertainty at such times would correspond physically to the opening of a window of opportunity that temporarily screens off complete determination by influences of the past.

When we do decide under such conditions of uncertainty, the outcome would not be determined because of the preceding indeterminacy—and yet it could be willed (and hence rational and voluntary) either way owing to the fact that, in such self-formation, the agents' prior wills are divided by conflicting motives. Consider an example I have commonly used of a businesswoman who faces a conflict of this kind. She is on the way to a meeting important to her career when she observes an assault taking place in an alley. An inner struggle ensues between her moral conscience, to stop and call for help, and her career ambitions which tell her she cannot miss this meeting. She has to make an effort of will to overcome the temptation to go on to her meeting. If she overcomes this temptation, it will

---

[5] That chaos plays a role in the activity of the brain is acknowledged by many: see, e.g., Skarda, and Freeman 1987; also Roskies (in press) and Walter 2001. Of course, chaotic behavior in itself is consistent with an underlying determinism, as is well known. But chaotic activity in the brain, because of sensitivity to initial conditions, could in principle amplify quantum indeterminacies at the synaptic or neuronal levels so that they would have non-negligible effects on neural processing.

be the result of her effort, but if she fails, it will be because she did not *allow* her effort to succeed. And this is due to the fact that, while she wanted to overcome temptation, she also wanted to fail, for quite different and competing reasons. When agents, like the woman, decide in such circumstances, and the indeterminate efforts they are making become determinate choices, they *make* one set of competing reasons or motives prevail over the others then and there *by deciding*.

Now add a further step. Just as, in this way, indeterminism need not undermine rationality and voluntariness, so indeterminism, in and of itself, need not undermine control and responsibility. Suppose you are trying to think through a difficult problem, say a mathematical problem, and there is some indeterminacy in your neural processes complicating the task—a kind of chaotic background. It would be like trying to concentrate and solve a problem with background noise or distraction. Whether you are going to succeed in solving the problem is uncertain and undetermined because of the distracting indeterministic noise. Yet, if you concentrate and solve the problem nonetheless, we have reason to say you did it and are responsible for it even though it was undetermined whether you would succeed. The distracting neural noise would have been an obstacle that you overcame by your effort.

There are numerous examples supporting this point (first suggested by Austin 1961, Anscombe 1971, and others) where indeterminism functions as an obstacle to success without precluding responsibility. Consider an assassin attempting to kill a prime minister with a high-powered rifle. He might fail because of undetermined events in his nervous system that might lead to a wavering of his arm. But if he does succeed nonetheless, despite this indeterminism, can he be held responsible? The answer is obviously yes because he voluntarily and intentionally succeeded in doing what he was *trying* to do—kill the prime minister. Yet his action, killing the prime minister, was undetermined. One might even say "he got lucky" in killing the prime minister, since there was a chance he might have missed. Yet, for all that, he *did* kill the prime minister and *was* responsible for it.

Here is another example I have used: A husband, while arguing with his wife swings his arm down on her favorite glass table-top, intending to break it. Again, we suppose some indeterminism in the nerves of his arm makes the momentum of his swing indeterminate so that it is literally not determined whether the table will break right up to the moment when it is struck. Whether the husband breaks the table or not is undetermined and yet he is clearly responsible if he does break it. (It would be a poor excuse for him to say to his wife "chance did it, not me". Though there was a chance he wouldn't break it, chance didn't do it, he did).

Such examples, to be sure, do not amount to genuine exercises of free will in SFAs, such as the businesswoman's, where the wills of the agents are divided between conflicting motives. The businesswoman wants to help the victim, but she also wants to go on to her meeting. By contrast, the will of the assassin is not equally divided. He wants to kill the prime minister, but does not also want to fail. (Thus, if he fails, it is *merely* by chance.) Yet these examples, of the assassin, husband and the like do provide some clues. To go further we have to add further steps.

6 Parallel Processing

Imagine in cases of conflict characteristic of SFAs, like the businesswoman's, that the indeterministic noise which is providing an obstacle to her overcoming temptation (and doing the moral thing) is not coming from an external source, but has its source in her own will, since she also deeply desires to do the opposite (go on to her meeting). To understand how this could be, imagine that two competing recurrent neural networks are involved, each influencing the other and representing her conflicting motivations.[6] The input of one of these networks consists in the woman's desires and motives for stopping to help the victim. If the network reaches a certain activation threshold (the simultaneous firing of a complex set of "output" neurons), that would represent her choice to help. For the competing network, the inputs are her ambitious motives for going on to her meeting and its reaching an activation threshold represents the choice to go on.

Now imagine further that the two networks are connected so that the indeterminism that is an obstacle to her making one of the choices is present because of her simultaneous conflicting desire to make the other choice—the indeterminism thus arising from a tension-creating conflict in the will, as noted. Under such circumstances, when either of the pathways reaches an activation threshold which amounts to choice, it would be like your solving the mathematical problem by overcoming the indeterministic background noise generated by the presence of the other pathway. And just as when you solved the mathematical problem despite the presence of this indeterminism, one could say you did it and are responsible for it, so one can say this as well, I would argue, in the present case, *whichever one is chosen*. The network through which she succeeds in reaching a choice threshold will have succeeded despite the indeterminism that was present because of the existence of the competing network.

---

[6] Recurrent networks are complex networks of interconnected neurons in the brain circulating impulses in feedback loops of a kind generally involved in higher level cognitive processing. An accessible introduction to the role of neural networks in cognitive processing which brings out the relevant features of recurrent networks is Spitzer 1999.

Note that, in these circumstances, the choices either way would not be "inadvertent", "accidental", "capricious", or "merely random", because they would be *willed* by the woman either way, when they are made, and done for *reasons* either way (moral convictions if she turns back, ambitious motives if she goes on) which she then and there endorses. And these are the conditions usually required to say something is done "on purpose", rather than accidentally, capriciously or merely by chance. Moreover, if we also assume (as we consistently can in the woman's case) that the agent is not being coerced (no one is holding a gun to her head), nor physically constrained or disabled, nor forced or controlled by others, then these conditions (that she wills it, does it for reasons, and could have done otherwise willingly and for reasons), rule out each of the normal reasons we have for saying that agents act, but do not have control over their actions (coercion, constraint, incapacity, inadvertence, involuntariness, mistake, or control by others).

To be sure, with "self-forming" choices of these kinds, agents cannot determine which choice outcome will occur *before* it occurs or the outcomes would be predetermined after all. But it does not follow that, because one does not determine which of a set of outcomes is going to occur before it occurs, one does not determine which of them occurs, *when* it occurs. When the above conditions for self-forming choices are satisfied, agents determine their future lives *then and there* by deciding.

Indeed, they have what I have called *plural voluntary control* over their options in the following sense: They are able to bring about *whichever* of the options they will, *when* they will to do so, for the *reasons* they will to do so, *on purpose* rather than by mistake or accident, without being coerced or compelled in doing so, or otherwise controlled by other agents or mechanisms. Each of these conditions can be satisfied in cases like the businesswoman's, whichever choice she makes, despite the indeterminism involved, as I have shown in many writings (see, e.g., Kane 1996, Chapters 8-10; 1999; 2005, Chapter 12). Satisfying them amounts in common parlance to the claim that the agents can choose either way "*at will*".

Note also that this account of self-forming choices amounts to a kind of "doubling" of the mathematical problem. It is as if an agent faced with such a choice is *trying* or endeavoring to solve two cognitive problems at once, or to complete two competing (deliberative) tasks at once—in our example, to make a moral choice and to make a conflicting self-interested choice (corresponding to the two competing neural networks involved). Each task is being thwarted by the indeterminism created by the presence of the other, so it might fail. But if it succeeds, then the agents can be held responsible because, as in the case of solving the mathematical problem, they will have succeeded in doing what they were trying or endeavoring to do. Recall the assassin and the husband once again. Owing to indeterminacies in their neural pathways, the assassin might miss his target or the husband fail to break the table. But if they *succeed*, despite the probability of failure, they are responsible, because they will have succeeded in doing what they were trying or endeavoring to do.

And so it is, I suggest, with self forming choices, except in their case, *whichever way the agents choose*, they will have succeeded in doing what they were trying to do because they were simultaneously trying to make both choices, and one is going to succeed. *Their failure to do one thing is not a mere failure, but a voluntary succeeding in doing the other.* Does it make sense to talk about agents trying to do two competing things at once in this way, or to solve two cognitive problems at once? Well, we know that the brain is a parallel processor; it can simultaneously process different kinds of information relevant to tasks such as perception or recognition through different neural pathways. Such a capacity, I believe, is essential to the exercise of free will.

In cases of self-formation (SFAs), agents are simultaneously trying to resolve plural and competing cognitive tasks. They are, as we say, of two minds. Yet they are not two separate persons. They are not dissociated from either task. The businesswoman who wants to go back to help the victim is the same ambitious woman who wants to go to her meeting and make a sale. She is a complex creature, torn inside by different visions of who she is and what she wants to be, as we all are from time to time. But this is the kind of complexity needed for genuine self-formation and free will. And when she succeeds in doing one of the things she is trying to do, she will endorse that as *her* resolution of the conflict in her will, voluntarily and intentionally, not by accident or mistake.

Note here, *contra* Frankfurt, the role of ambivalence or conflict in the will in all this. If we were never ambivalent or moved by conflicting values, multiple valences, we could never be "self forming" beings. Of course it is also true that if we were *always* ambivalent we would never be self *formed* beings.

7 Objections and Responses (I): Introspection, Efforts, Rationality and Will-setting

Needless to say, there are many questions and objections that may be raised about this view that I have tried to address in many writings (Kane 1996, 1999; 2005. See also 2002). In the remainder of this paper, I will address some of the more important of these objections.

A frequently-made objection is that we are not introspectively or consciously aware of making dual efforts and performing multiple cognitive tasks in self-forming choice situations. But I am not claiming that agents are introspectively aware of making dual efforts. What persons are introspectively aware of in SFA situations is that they are trying to decide about which of two options to choose and that either choice is a difficult one because there are resistant motives pulling them in different directions that will have to be overcome, *whichever* choice is made. In such introspective conditions, I am theorizing that what is going on underneath is a kind of distributed processing in the brain that involves separate attempts or endeavorings to resolve competing cognitive tasks.

There is a larger point here I have often emphasized: *Introspective evidence cannot give us the whole story about free will.* Stay on the introspective surface and libertarian free will *is* likely to appear obscure or mysterious, *as it so often has in history*. What is needed is a *theory* about what might be going on behind the scenes when we exercise such a free will, not merely a description of what we immediately experience; and in this regard new scientific ideas can be a help rather than a hindrance to making sense of free will. It is now widely believed that parallel processing takes place in the brain in such cognitive phenomena as visual perception. The theory is that the brain separately processes different features of the visual scene, such as object and background, through distributed and parallel, though interacting, neural pathways or streams.

Suppose someone objected that we are not introspectively aware of such distributed processing in ordinary cases of perception. That would hardly be a decisive objection against this new theory of vision. For the claim is that this is what we are doing in visual perception, not necessarily that we are introspectively aware of doing it. And I am making a similar claim about free will. *If parallel distributed processing takes place on the input side of the cognitive ledger (in perception), then why not consider that it also takes place on the output side (in practical reasoning, choice and action)?* That is what I am suggesting we should suppose if we are to make sense of libertarian free will.

Another commonly made objected is that it is irrational to make efforts to do incompatible things. I concede that in most ordinary situations it is. But I contend that there are special circumstances in which it is not irrational to make competing efforts: These include circumstances in which (i) we are deliberating between competing options; (ii) we intend to choose one or the other, but cannot choose both; (iii) we have powerful motives for wanting to choose each of the options for different and competing reasons; (iv) there is a consequent resistance in our will to either choice, so that (v) if either choice is to have a chance of being made, effort will have to be made to overcome the temptation to make the other choice; and most importantly, (vi) we want to give each choice a fighting chance of being made because the motives for each choice are important to us. The motives for each choice define in part what sort of person we are; and we would taking them lightly if we did not make an effort in their behalf. These are the conditions of "will-setting" or "self-forming" actions (SFAs).

It is critical here to recognize the uniqueness of such "will-setting" situations. For our normal intuitions about efforts are formed in everyday situations in which our will is already "set one way" on doing something, where obstacles and resistance have to be overcome if we are to succeed in doing it. We want to open a door, which is jammed, so we have to make an effort to open it. In such everyday situations, it would be irrational to make incompatible efforts because our wills are already set on doing what we are trying to do. In will-setting situations by contrast, one's will is not yet set on doing either of the things one is trying to do, though one has strong reasons for doing each, neither of which are as yet *decisive*. Because most efforts in everyday life are made in will-settled situations where our will is already set on doing what we are trying to do, we tend to assimilate all effort-making to such situations, thereby failing to consider the uniqueness of will-setting, which is of a piece, in my view, with the uniqueness of *free will*.

8 Objections and Responses (II): Chance, Luck, Indeterminism, Control and Action

Perhaps the most important objections, however, concern luck and chance, which have taken many forms in contemporary debates about free will. Here is a particularly powerful form such objections have taken: If the occurrence of a choice depends on the occurrence of some undetermined or chance events (say, quantum events) in the brain over which the agent lacks control, then whether or not the choice occurs would appear to be a matter of luck or chance, rather than something the agent brought about and was responsible for.

Such thoughts, as noted, tend to send one looking for extra factors, other than prior events or happenings, to "tip the balance" to one choice or the other. But there is an alternative way to think about how indeterminism might be involved in free choice that first occurred to me thirty years ago, a way that avoids such familiar stratagems and requires a transformation of perspective.

Think, instead, of the indeterminism involved in free choice as an *ingredient* in larger goal-directed or teleological processes or activities, in which the indeterminism functions as a *hindrance* or *obstacle* to the attainment of the goal. Such is the role of indeterminism in the efforts or "volitional streams", as I sometimes call them, preceding undetermined SFAs. Each of these streams is a temporally extended goal-directed activity, whose goal is a particular choice and whose input consists in the reasons or motives for making that choice, in which indeterminism is a hindering or interfering element. The choices or SFAs that result from these temporally extended activities, thus do not pop up out of nowhere, even though undetermined. They are the *achievements* of goal-directed activities of the agent that might have failed due to the indeterminism, but did not.

Note that if indeterminism or chance does play this kind of interfering role in larger goal-directed processes leading to choice, the indeterminism or chance needs not be the *cause* of the choice that is actually made. This follows from a general point about probabilistic causation. A vaccination may hinder or lower the probability that I will get a certain disease, so it is causally relevant to the outcome. But if I get the disease despite it, the vaccination is not the *cause* of my getting the disease, though it was causally relevant, because its role was to *hinder* that effect. The causes of my getting the disease, by contrast, are those causally relevant factors (such as the infecting virus) that significantly enhanced the probability of its occurrence.

Similarly, in the businesswoman's case, the causes of the choice she does make (the moral choice or the ambitious choice) are those causally relevant factors that significantly *raised* the probability of making *that* choice from what it would have been if those factors had not been present. Such factors would include her reasons and motives for making that choice rather than the other, her conscious awareness of these reasons and her deliberative efforts to overcome the temptations to make the contrary choice. The indeterminism or chance involved (like the vaccination) was causally relevant to the outcome, but it was not the cause. This explains why the husband's excuse was so lame when he said "Chance broke the table, not me". The chance was a hindering factor, not the cause.

Another common set of objections concern the crucial notion of *control*. If free choices are undetermined, it would seem that the agent cannot have control over which one occurs. But again, this is one of those intuitions that seems obvious, even undeniable, at first blush, but turns out not to be so. For an agent to have *control* at a time t over the being or not being of some event (such as a choice) at that time is for the agent to have the *ability* or *power* at t to *make* that event *be* at t *and* the ability or power at t to make it *not be* at t. And in an SFA, an agent exercises just this kind of control over the choice that is made at the time it is made. For the agent not only had the ability or power at that time to make that choice be, the agent also had the ability or power at that time to make it not be, *by making the competing choice be*.

Not only would the agent have control over both choices in a SFA situation in this sense, but the control would be what was earlier called *plural voluntary control*. For as explained earlier, the agent would not only have had the power to make either choice be or not be at the time, but would have had the power to do this either way *voluntarily, on purpose and for reasons*, and not merely by accident; and this would be so *even though* the occurrence of the choice that did occur depended on the firing or non-firing of neurons over which the agent had no control. An astonishing outcome! But it is the result one gets by (1) giving indeterminism an interfering or hindering role in larger goal-directed activities and (2) allowing for multiple such activities or volitional streams in deliberation.

Perhaps the strongest of objections against this view, however, are so-called "luck objections", which have taken explanatory forms. The most influential and the strongest of these I believe is that given by Alfred Mele (1998): "If different free choices could emerge from the same past of an agent, there would seem to be no explanation for why one choice was made rather than another in terms of the total prior character, motives and purposes of the agent. The difference in choice, i.e. the agent's choosing one thing rather than another, would therefore be just a matter of luck".

On the model I have given it is true enough that different choices could emerge from the total prior character, motives and purposes of the agent. But it does not follow that the choice made is "just a matter of luck". First, on that model, whichever choice is made, the agent causes or brings about the choice by engaging in a goal-directed process of trying or attempting to bring about that very choice and succeeding in attaining the goal. Second, the agent has control over the occurrence of the choice that is made when it is made in the sense of control just defined. Third, it does not follow that the choice was irrational since it was made for reasons that the agent then and there endorsed as the reasons he or she would act on. Fourth, the choice was made voluntarily either way in the sense of not being coerced or compelled or otherwise constrained. Fifth, the choice was made on purpose rather than by mistake or accident either way.

Finally, as a consequence of all these things, that the agent brought about the choice, had control over it, did so voluntarily (without being coerced or compelled), knowingly and on purpose, and for reasons, and could have done otherwise voluntarily, knowingly, on purpose and for reasons, one can say that the agent was responsible for the outcome just as one can say this of the assassin when he succeeded in killing the Prime Minister or the husband when he succeeded in breaking the table. If the conclusion of this luck objection—namely, that the choices made are "just a matter of luck"—is meant to be consistent with all this (the agent brought it about, had control over it, acted voluntarily, knowingly, on purpose, etc.) then the conclusion, and the objection itself, would lose all traction.

9 Liberum Arbitrium Voluntatis

Well, not quite all traction. And this is where things get interesting. With powerful arguments in philosophy, it is not enough to show that their conclusions do not necessarily follow from their premises. One needs also to show why the objections seem to have such power and seem irrefutable. In the case of this explanatory luck objection I think this is because it does show something of very great importance about free will. It shows that there is something to the often repeated charge that undetermined choices would be *arbitrary* in a certain sense. A residual arbitrariness seems to remain in all self-forming choices since the agents cannot in principle have sufficient or overriding ("conclusive" or "decisive") *prior* reasons for making one option and one set of reasons prevail over the other.

Therein lies the truth in this luck objection: a free choice cannot be *completely explained* by the entire past, including past causes *or reasons*; and I think this is a truth that reveals something important about free will. I have argued elsewhere that such arbitrariness relative to prior reasons tells us that every undetermined self-forming choice is the initiation of novel pathway into the future, whose justification lies in that future and is not fully explained by the past. In making such a choice we say, in effect, "I am opting for this pathway. It is not *required* by my past reasons, but is consistent with my past and is one branching pathway my life can now meaningfully take. Whether it is the right choice, only time will tell. Meanwhile, I am willing to take responsibility for it one way or the other".

Of special interest here, as I have often noted in my writings, is that the term "arbitrary" comes from the Latin *arbitrium*, which means "judgment"—as in *liberum arbitrium voluntatis*, "free judgment of the will" (the medieval designation for free will). Imagine a writer in the middle of a novel. The novel's heroine faces a crisis and the writer has not yet developed her character in sufficient detail to say exactly how she will act. The author makes a "judgment" about this that is not determined by the heroine's already formed past which does not give unique direction. In this sense, the judgment (*arbitrium*) of how she will react is "arbitrary", but not entirely so. It had input from the heroine's fictional past and in turn gave input to her projected future.

In a similar way, agents who exercise free will are both authors of and characters in their own stories at once. By virtue of "self-forming" judgments of the will (*arbitria voluntatis*) (SFAs), they are "arbiters" of their own lives, "making themselves" out of past that, if they are truly free, does not limit their future pathways to one. If we should charge them with not having sufficient or *conclusive* prior reasons for choosing as they did, they might reply: "True enough. But I did have *good* reasons for choosing as I did, which I'm willing to endorse and take responsibility for. If they were not sufficient or conclusive reasons, that's because, like the heroine of the novel, I was not a fully formed person before I chose (and still am not, for that matter). Like the author of the novel, I am in the process of writing an unfinished story and forming an unfinished character who, in my case, is myself".

References

Anscombe, G.E.M 1971, *Causality and Determinism*. Cambridge: Cambridge University Press.
Aristotle 1915, *Nichomachean Ethics*, in W.D. Ross (ed.), *The Works of Aristotle,* Vol. 9, London: Oxford University Press.
Atmanspacher, H., and Rotter, S. 2008. "Interpreting Neurodynamics: Concepts and Facts", *Cognitive Neurodynamics* 2: 297-318.
Austin, J.L. 1961, "Ifs and Cans", in *Philosophical Papers*, ed. by J. Urmson and P. Warnock, Oxford: Clarendon Press, 153–180.
Balaguer, M. 2010, *Free Will as an Open Scientific Problem*, Cambridge MA: MIT Press.
Beck, F. and Eccles, J. 1992, "Quantum Aspects of Brain Activity and the Role of Consciousness", *Proceedings of the National Academy of Science USA* 89: 11357-11361.
Brembs B. 2010, "Towards a Scientific Concept of Free Will as a Biological Trait: spontaneous actions and decision-making in invertebrates", *Proceedings of the Royal Society B: Biological Sciences* 278 (1707): 930–939. DOI: 10.1098/rspb.2325.
Frankfurt, H. 1971, "Freedom of the Will and the Concept of a Person", *Journal of Philosophy* 68: 5–20.
Glimcher, P.W. 2005, "Indeterminacy in Brain and Behavior", *Annual Review of Psychology* 56: 25–56.
Heisenberg, M. 2013, "The Origin of Freedom in Animal Behavior", in A. Suarez and P. Adams (eds.), *Is Science Compatible with Free Will?*, Dordrecht: Springer: 95–103.
Kane, R. 1996, *The Significance of Free Will*, Oxford: Oxford University Press.
Kane, R. 1999, "Responsibility, Luck and Chance: Reflections on Free Will and Indeterminism", *Journal of Philosophy* 96: 217–40.
Kane, R. 2002, "Some Neglected Pathways in the Free Will Labyrinth", in R. Kane (ed.), *The Oxford Handbook of Free Will*, 1st Edition, Oxford: Oxford University Press, 406–437.
Kane, R. 2005, *A Contemporary Introduction to Free Will*, Oxford: Oxford University Press.
Mele, A. 1998, Review of Robert Kane, *The Significance of Free Will* (Oxford: Oxford University Press, 1996), *Journal of Philosophy* 95: 581–584.
Roskies, A. (in press), "Can Neuroscience Resolve Issues about Free Will?", forthcoming in W. Sinnott-Armstrong (ed.), *Moral Psychology: Volume 4*, Cambridge MA: MIT Press.
Penrose, R. 1994. *Shadows of the Mind*. Oxford: Oxford University Press.
Skarda, C., and Freeman, W. 1987, "How the Brain Makes Chaos in Order to Understand the World", *Behavioral and Brain Sciences*, 10: 161–195.
Spitzer M. 1999, *The Mind Within the Net*, Cambridge MA: MIT Press.
Strawson, P.F. 1962, "Freedom and Resentment", *Proceedings of the British Academy* 48: 1–25.
Walter, H. 2001, *Neurophilosophy of Free Will*, Cambridge MA: MIT Press.

Peter Jedlicka
Quantum Stochasticity and (the End of) Neurodeterminism

1 Introduction

Free will is a "scandal in philosophy" (Doyle 2011). However, free will is no scandal for contemporary mainstream neuroscience. Most neurobiologists believe that there is no such problem or that neuroscience has already solved the problem of free will or is on the way to solving it by fully elucidating the neuronal mechanisms of human decisions. A completely mechanistic explanation of "free" decisions would make them causally dependent on neuronal activity and thus not free. Hence most neuroscientists consider free will an illusion (e.g. Wegner 2003) or they adopt a compatibilist view, believing that free will can exist even in the fully mechanistic machine called the human brain. Many neurobiologists would agree with Francis Crick who famously said that

> "You", your joys and your sorrows, your memories and your ambitions, your sense of personal identity and free will, are in fact no more than the behavior of a vast assembly of nerve cells and their associated molecules (Crick 1994, 3).

Although new ideas and concepts are emerging (e.g. Van Regenmortel 2004, Noble 2006), reductionism and determinism are still the major paradigms in current biology, including neurobiology. "[Physicists] invented the deterministic-reductionistic philosophy and taught it to the biologists, only to walk from it themselves" (Loewenstein 2013). The dominant belief is that "anything can be reduced to simple, obvious mechanical interactions. The cell is a machine. The animal is a machine. Man is a machine" (Monod 1974, IX).[1] In this article, I am going to argue that the mainstream Newtonian view of the human brain as a sophisticated machine is challenged by new findings in neuroscience and in the rising field of quantum biology.

---

[1] Unfortunately, scientists are often not aware that their fully deterministic and reductionistic view of the world is a philosophical (metaphysical) view, and not a scientific view based on purely empirical or logical evidence (cf. Jedlicka 2005).

2 Definition of neurodeterminism

In the early 19th century, inspired by Newtonian physics, Pierre-Simon Laplace expressed the deterministic world view in his famous description of what is now known as *Laplace's Demon*: an intelligent being that knows completely the state of the world (all forces acting in the universe and the position of all particles) at one moment and can calculate its whole future development:

> We ought to regard the present state of the universe as the effect of its antecedent state and as the cause of the state that is to follow. An intelligence knowing all the forces acting in nature at a given instant, as well as the momentary positions of all things in the universe, would be able to comprehend in one single formula the motions of the largest bodies as well as the lightest atoms in the world, provided that its intellect were sufficiently powerful to subject all data to analysis; to it nothing would be uncertain, the future as well as the past would be present to its eyes (Laplace 1814, 19).

Following Laplace, we can define a *deterministic system* as a system whose behavior is constrained in such a ways that its inputs and initial state fully determine its next state or output. In other words, "[t]he world is governed by (or is under the sway of) determinism if and only if, given a specified way things are at a time t, the way things go thereafter is fixed as a matter of natural law" (Hoefer 2010). On the contrary, a stochastic or indeterministic system can be defined as a system whose inputs and initial state do not fully determine its next state or output. Expressed in a more formal way, if S is a complete description of the state of our world at a given time t and L is a complete description of all laws of nature (e.g., the laws of Newtonian mechanics), then S and L together logically imply a complete description of the entire history of our world following t:

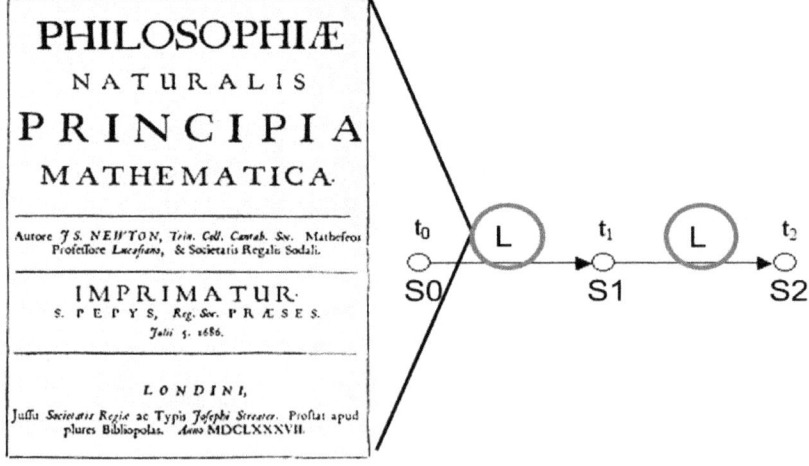

In such a deterministic world, all events are the necessary consequences of the previous states and the laws of the universe. For instance, in the deterministic scenario, the formation of the planet Earth and the extinction of dinosaurs would be simply an inevitable consequence of the state of the universe at the time of the Big Bang and of the deterministic laws of physics.

Neurodeterminism can be seen as a special case of cosmic determinism, because *SB*, the state of the brain and its environment (inputs) at a given time *t*, is a part of the state of the world *S* and thus is along with *S* fully governed by the antecedent states of the world and its laws *L*. This definition is similar to the description of neural causation by the neuroscientist Wolf Singer:

> If neuronal processes are the basis and cause of all mental phenomena and if brain processes follow the laws of nature, then the principle of causality must hold for neuronal interactions. Even though there is noise and interference, each state of the brain is then a necessary consequence of the immediately preceding state. Since decisions are the consequence of special brain states, our concepts concerning the independence of will are likely to require some revision (Singer 2009).

3 The incompatibility of neurodeterminism and responsibility

Are free will and moral responsibility compatible with (neuro)determinism? Many influential thinkers are compatibilists (Mckenna 2009), for example Daniel Dennett (2003) or Harry Frankfurt (1969). However, a strong argument has been raised against the compatibility of free will, moral responsibility and determinism. It is called the consequence argument:

> If determinism is true, then our acts are the consequence of laws of nature and events in the remote past. But it's not up to us what went on before we were born, and neither is it up to us what the laws of nature are. Therefore, the consequences of these things (including our present acts) are not up to us (Van Inwangen 1983, 56).

If all our acts are unavoidable consequences of the laws of nature and of events before our birth then we are not free to choose between alternative courses of action. We cannot do otherwise. Therefore, we have no moral responsibility for our actions. This is an indirect consequence argument (Jäger 2006) for the incompatibility of moral responsibility and determinism because it relies on (1) the incompatibility of free will and determinism ("free will incompatibilism") and (2) the principle of alternative possibilities (PAP). PAP states that we are morally responsible for what we have done only if we could have done otherwise (Frankfurt 1969). Using PAP it is possible to make the logical step from "free will incompatibilism" to "moral responsibility incompatibilism". However, PAP has been attacked by Harry Frankfurt and other compatibilist (Frankfurt 1969, see also McKenna 2009). Nevertheless, it is possible to construct direct consequence arguments for the incompatibility of moral responsibility and determinism which do not depend on the validity of PAP. Christoph Jäger has recently formulated three versions of such a direct consequence argument (Jäger 2006). The strongest version can be formulated in nontechnical terms as follows:

$S0$ is the description of the state of the universe as it was at a time point before human beings appeared in the history of evolution. L is the description of the laws of the universe, and A is the description of some arbitrarily chosen human action that happened (or is going to happen). The argument consists of one inference rule ("Boethius' principle") and three steps. Boethius' principle is the following: From "p and no one is responsible for the fact that p, and it is necessary that p implies q" it can be inferred: "q and no one is responsible for the fact that q". The rule is valid for all p and q in a deterministic universe in which all human cognitive acts (desires, beliefs) and behavioral actions are fully determined. The reformed consequence argument draws a devastating conclusion from two premises and Boethius' principle:
1. $S0$ and L, and no one is responsible for the fact that $S0$ and L (premise 1)
2. It is necessary that $S0$ and L imply A (premise 2, which follows from determinism)
3. A and no one is responsible for the fact that A (conclusion, based on 1, 2 and Boethius' principle)

The consequence argument is powerful because premise 2 is based on determinism and premise 1 is difficult to refute.[2] So the only plausible way of criticism is to attack the inference rule.

The present version of the consequence argument has three important advantages over previous versions of the consequence arguments: (1) it does not require PAP, (2) Boethius' principle is immune against the usual criticism of the inference rules used in consequence arguments[3] and (3) Boethius' rule makes the argument also immune against counterexamples based on overdetermination (Jäger 2006).

---

[2] But see an interesting rejection of premise 1 based on multiple-pasts compatibilism in Aaronson 2013, 24.
[3] For a discussion of the validity of the inference rule β, see e.g. Vihvelin 2011.

In sum, the consequence argument shows that compatibilists do not have many plausible options defending the compatibility of determinism and moral responsibility.[4] And since neurodeterminism is just a special case of cosmic determinism, the consequence argument denies also the compatibility of neurodeterminism and moral responsibility. If our universe were a machine governed by deterministic laws and initial conditions, then our brain would be also a machine (a computer), and, consequently our experience of conscious will and moral responsibility would be only an illusion.

4 The exorcism of Laplace's Demon by quantum theory

If determinism is incompatible with moral responsibility, then the most important empirical question is the following: Is our world deterministic? There was not much hope for free will or moral responsibility in the times of Newtonian physics. But then came the quantum revolution. Theories and experiments in quantum physics strongly support the indeterministic nature of our universe. Quantum theory is indeterministic. The development of the quantum state of a physical system is governed by the deterministic wave function. However, according to most interpretations of quantum mechanics, the result of a quantum measurement (which corresponds to a collapse of the wave function) is undetermined. Thus, quantum measurements generate objectively indeterminate events: their cause is nothing or something outside the physical world (Satinover 2001, 101). Even with full knowledge of the state of a quantum mechanical system, it is not possible to predict (calculate) the result of a measurement, only its probability. There is no underlying mechanistic cause (within the physical world) for the specific actual outcome of a measurement when there are several possible outcomes.

There are two alternative options for a deterministic interpretation of quantum theory.[5] But both have their weak points. (1) The many-worlds interpretation has

---

[4] Is indeterminism compatible with moral responsibility and free will? Indeterminism alone is not sufficient for responsibility and free will, but the consequence argument shows that indeterminism is a necessary condition (*a conditio sine qua non*) for responsibility and free will. A fully indeterministic world with no deterministic laws would not be compatible with responsibility, since in a purely random and chaotic universe we would not have any control over our acts. Thus, a universe with a mixture of partial determinism and partial indeterminism is required for moral responsibility.

[5] Interpretations of quantum mechanics can be classified into three groups depending on which

serious troubles with probability. According to the many-world interpretation, all outcomes are actualized. But "[o]nce you give up the distinction between actuality and possibility saying that all possible realities are equally actual, the notion of probability becomes meaningless" (Putnam 2005). (2) The most popular hidden variable theory is Bohmian mechanics which "buys its 'determinism' via the mathematical device of pushing all the randomness back to the beginning of time" (Aaronson 2013, 23). In sum, we can say with Scott Aaronson that

> [P]hysical indeterminism is now a settled fact to roughly the same extent as evolution, heliocentrism, or any other discovery in science. So if that fact is considered relevant to the free-will debate, then all sides might as well just accept it and move on! (2013, 23)

The quantum revolution in the direction of indeterminism was shocking to many physicists. Einstein famously said that God does not play dice. However, it seems that God plays dice after all. Quantum theory has performed a successful exorcism on Laplace's Demon. And there are good reasons to expect that future physical theories will also contain quantum indeterminism (e.g. Banks 1997). Quantum indeterminacy is relevant to the problem of free will because it has swept away Newtonian physical determinism (Barr 2003). Of course, quantum theory has not proven that free will exists. However, it has shown that the classical determinism of Laplace can no longer be plausibly used as an argument from physics against free will. Thus, quantum theory opened new ways to reconcile human freedom with the laws of physics (Weyl 1932).

5 Two arguments against the quantum brain hypothesis

Quantum physics has shaken classical determinism. However, whether and how it can shake neurodeterminism is far from clear and straightforward. It is true that the only objectively indeterministic processes in the physical world are quantum processes. But does quantum indeterminacy affect the dynamics of neuronal networks? Does quantum physics allow the brain to exploit indeterminate quantum

---

the following three assumptions is denied (Brüntrup 2008, 60): 1. Development of a physical system in time is fully governed by the wave function (→ collapse-of-the-wave-function theories). 2. There are no hidden variables (→ hidden variable theories). 3. Measurements always provide a single definite outcome (→ many-worlds interpretation).

events? These are key empirical questions. There are two main arguments which are usually raised against the quantum brain hypothesis:

1. Neuronal signaling molecules, neurons and neural networks are *too large* for quantum phenomena to play a significant role in their functioning. The conventional wisdom is that all *quantum events* are *averaging out*, so that fluctuations among quantum particles are not important. As expressed by Daniel Dennett:

> Most biologists think that quantum effects all just cancel out in the brain, that there's no reason to think they're harnessed in any way. Of course they're there; quantum effects are there in your car, your watch, and your computer. But most things—most macroscopic objects—are, as it were, oblivious to quantum effects. They don't amplify them; they don't hinge on them (quoted in Penrose 1996, 251).

Christoph Koch and Klaus Hepp (2006), too, identify the large size of neuronal objects and the huge number of particles involved in neuronal signaling as one of the critical weak points of quantum brain hypothesis:

> Although brains obey quantum mechanics, they do not seem to exploit any of its special features. Molecular machines, such as the light-amplifying components of photoreceptors, pre- and post-synaptic receptors and the voltage- and ligand-gated channel proteins that span cellular membranes and underpin neuronal excitability, are so large that they can be treated as classical objects. [...] Two key biophysical operations underlie information processing in the brain: chemical transmission across the synaptic cleft, and the generation of action potentials. These both involve thousands of ions and neurotransmitter molecules, coupled by diffusion or by the membrane potential that extends across tens of micrometres. Both processes will destroy any coherent quantum states. Thus, spiking neurons can only receive and send classical, rather than quantum, information. It follows that a neuron either spikes at a particular point in time or it does not, but is not in a superposition of spike and nonspike states.

2. The second important criticism is that the interaction of neuronal molecules, neurons or neuronal networks with their noisy, wet and warm environment will destroy any nontrivial quantum states such as superpositions or entanglements. If this were true, only trivial quantum effects could be present in the nervous system. But what is the difference between trivial and nontrivial quantum effects? Trivial quantum effects provide the basis for the structure and chemical properties of molecules and they are ubiquitous (also in cars and watches). Hence, trivial quantum effects are crucial for the basic biochemistry of neuronal molecules, but when considering the neuronal function, these trivial quantum features can, allegedly, be ignored and the molecules important for neuronal signaling can be treated as essentially classical. All interesting coherent quantum states which are necessary for any nontrivial quantum computation (Davies 2004) can, allegedly,

exist only in well isolated quantum states and are rapidly destroyed by the environment. Since the brain is "a 300-degrees Kelvin tissue strongly coupled to its environment" (Koch, and Hepp 2006), decoherence will prevail and no neuronal quantum computation will be possible. Because of the extremely high speed of environment-induced decoherence, the brain "should be thought of as a classical rather than quantum system" (Tegmark 2000).

6 Towards quantum neurobiology

The two arguments against the hypothesis that quantum dynamics play a nontrivial role in the nervous system seem convincing and are accepted by most scientists. However, a growing body of empirical evidence indicates that the second argument is false and that the first argument is also very likely false. Exciting recent research shows that *nontrivial quantum effects* are present *in biological systems*—and not just in spite of, but sometimes because of, the interaction with the noisy and warm environment. Furthermore, because the brain is a *complex nonlinear system* with high sensitivity to small fluctuations, it is likely that it can *amplify* microscopic *quantum effects*. Specifically, there are two alternative but interrelated ways in which quantum events may influence the activity of the brain (Satinover 2001, 210; see also Jedlicka 2009): (1) Nontrivial quantum effects can speed up the computational processes in living organisms at the microscopic level. (2) Nonlinear chaotic dynamics can amplify lowest-level quantum fluctuations upward, modulating even larger-scale macroscopic neuronal activity.

What is the experimental evidence for these two claims?
(1) Contrary to expectations, nontrivial quantum processes have been observed in living systems. Recent experiments provide evidence for unexpectedly long-lasting quantum coherence in the electron transfer which is involved in photosynthesis.[6] This quantum-mechanical process is thought to improve the efficiency of energy transfer in photopigment molecules (Panitchayangkoon, et al. 2010). The pigment molecules seem to implement an efficient quantum algorithm to find the fastest route for the light-induced excitation of electrons (Sension 2007; but see also Tiersch et al. 2012). Quantum coherence has been found in photosynthetic bacteria as well as in marine algae. This suggests that evolution has been able to select and exploit quantum-mechanical features for fast and efficient computation in two evolutionary distinct organisms. Another

---

[6] Engel et al. 2007; Lee et al. 2007; Collini et al. 2010; Panitchayangkoon et al. 2010; Sarovar et al. 2010.

example of quantum dynamics in living systems has been found in photoreceptors, which are important for vision. Photoreceptor cells of the retina contain a protein called rhodopsin. Experiments using high-resolution spectroscopic and nuclear-magnetic resonance techniques revealed coherent quantum waves in the rhodopsin molecule (Wang et al. 1994; Loewenstein 2013). As summarized by Werner Loewenstein (2013):

> Quantum mechanics, not classical mechanics, rules the roost at this sensory outpost of the brain.

Quantum effects have been described also in the olfactory system. Electron tunneling has been suggested to play an important role in the detection of odorants by olfactory receptors (Huelga, and Plenio 2013). Avian magnetoreception is yet another example of potentially beneficial quantum effects in biology. Long-lived quantum entanglements in the cryptochromes of the retina seem to support the sensitivity of a bird's eye to magnetic fields (Huelga, and Plenio, 2013; Arndt et al. 2009; Ball 2011). In addition, quantum tunneling has been observed in other biomolecules, such as enzymes or motor proteins (Hunter 2006). Most importantly, contrary to the long-held view, under some conditions, the strong coupling to the noisy and warm environment is able to promote rather than hinder long-lasting quantum coherence in biological systems (Plenio, and Huelga 2008; Huelga, and Plenio 2013). Because of the accumulating evidence that quantum phenomena need to be considered explicitly and in detail when studying living organisms, quantum biology has recently emerged as a new field at the border between quantum physics and the life sciences (Ball 2011).

> Physicists thought the bustle of living cells would blot out quantum phenomena. Now they find that cells can nurture these phenomena—and exploit them (Vedral 2011).

So far, we have focused on nontrivial quantum processes in sensory cells. But are nontrivial quantum effects also present elsewhere in the nervous system? It is very likely, but direct experimental evidence is still missing. Where should we look for further instances of neuronal quantum effects? There are many stochastic neuronal mechanisms which may be driven by quantum events. Although it is true that "the main sources of neural noise are forces that can be characterized as thermal and chaotic rather than quantum in nature" (Sompolinsky 2005, 31), quantum physics is expected to shape at least some stochastic events in the brain, such as the opening of ion channels (Vaziri, and Plenio 2010). In this way microscopic quantum events might affect electrical signals in neurons, as proposed by Paul Glimcher (2005):

> [T]hese data suggest that membrane voltage is the product of interactions at the atomic level, many of which are governed by quantum physics and thus are truly indeterminate events. Because of the tiny scale at which these processes operate, interactions between action potentials and transmitter release as well as interactions between transmitter molecules and postsynaptic receptors may be, and indeed seem likely to be, fundamentally indeterminate.

It is in itself important that quantum coherence in living organisms has been experimentally demonstrated at the microscopic level. But what are the spatial and temporal limits of these quantum effects? Can we discover quantum coherence in more than just a few molecules? How long can it persist? Some quantum brain proposals, focusing on explaining consciousness (Hameroff, and Penrose 2014), would require coherent quantum waves on much larger and longer scales than those found so far. The honest reaction to the questions just posed is that we do not know the answers to them and that only future research can provide those answers.

(2) If it turned out that quantum effects cannot be observed in living systems at the macroscopic level, would that mean that living systems can be fully described by classical physics? Or is there another plausible way in which small-scale quantum effects—there is evidence for their occurrence (see (1))—might influence large-scale neuronal activity and behavior? Yes, there is. The common view that minuscule fluctuations, including quantum events, cancel out in larger systems need not be true in highly nonlinear systems like our brain. The nervous system can be seen as a nested hierarchy of nonlinear complex networks of molecules, cells, microcircuits and brain regions. In iterative hierarchies with nonlinear dynamics (at the edge of chaos), small (even infinitesimal) fluctuations are not averaged out, but can be amplified. Quantum fluctuations on the lowest level of scale may influence the initial state of the next level of scale, while the higher levels shape the boundary conditions of the lower ones. This hierarchy of nested networks with many feedback loops exploits rather than cancels out the quantum effects:

> [Q]uantum dynamics alters the final outcomes of computation at all levels—not by producing classically impossible solutions but by having a profound effect on which of many possible solutions are actually selected (Satinover 2001, 210).

In his essay on free will and neuroscience, Haim Sompolinsky has also mentioned this possibility:

> Chaos within the brain may amplify enormously the small quantum fluctuations [...] to a degree that will affect the timing of spikes in neurons (2005, 32).

Similarly, even Christof Koch, one of the major critics of quantum brain ideas, had to admit:

> What cannot be ruled out is that tiny quantum fluctuations deep in the brain are amplified by deterministic chaos and will ultimately lead to behavioral choices (2009, 40).

The quantum amplification mechanism has been adopted also by Scott Aaronson, whose recent "freebit" theory of free will "postulates that chaotic dynamics in the brain can have the effect of amplifying freebits to macroscopic scale" (2013, 38).[7] A similar theory has already been proposed by Pascual Jordan (1938).

What is the evidence for the proposal that the brain is a complex nonlinear system, capable of chaotic dynamics? Beggs and Plenz (2003, 2004) provided experimental evidence that neuronal networks can produce complex patterns of collective activity, which are called *neuronal avalanches.* These avalanches have a characteristic distribution: Each avalanche engages a variable number of neurons, but, on average, many more small avalanches are observed than large ones. This indicates that neuronal networks are poised *near criticality* (near phase transition, see Beggs and Timme 2012) and are prone to displaying emergent *complex* activity (Chialvo 2010). Similar results supporting *criticality* in the brain have been obtained on a larger scale from fMRI data (e.g. Deco, and Jirsa 2012). In general, we can observe three types of dynamics in the brain: 1. *ordered/subcritical* dynamics consisting of oscillatory synchronous activity with the characteristic features of high coordination and low variability, 2. *random/supracritical* dynamics consisting of asynchronous irregular activity with low coordination and high variability, and 3. *complex/critical* dynamics with high coordination and high variability. Brain states exhibiting *complex/critical* dynamics are the most interesting ones because they support the most efficient information processing (Beggs and Timme 2012). At the critical point between order and disorder (i.e. at the edge of chaos and instability), neurons can communicate best, since at that point they are coordinated but not stuck in a certain state for a long time and can establish long-range dynamical correlations. Furthermore, neuronal networks in near critical states display, because of the largest fluctuations, the largest repertoire of network activity. Finally, at the critical point, the highest sensitivity to small fluctuations (e.g. London et al. 2006) is observed: even a single neuron perturbation has a small but non-zero chance to trigger an avalanche. As pointed out by Dante Chialvo (2010), there are convincing Darwinian reasons for supposing that (parts of) our brains operate near the critical point: In a *subcritical* world, everything would always be uniform, there would be nothing new to learn and hence no critical and plastic brain would be needed; memories might as well be unchanging. In a *supracritical* world, everything would always be changing with

---

[7] But see also criticism of this quantum amplification idea in Clarke 2014.

no regularities to be learnt. No long-term plasticity and memory would be of any help. In our *critical* (complex) world, surprising events do occur, but regularities, too, are present so that the brain needs to register but also to update the stored memories.

> [B]rains seem "balanced on a knife-edge" between order and chaos: were they as orderly as a pendulum, they couldn't support interesting behavior; were they as chaotic as the weather, they couldn't support rationality (Aaronson 2013, 48).

Thus, it is highly plausible that small quantum fluctuations can be amplified, since brain activity can develop to the critical point: the point of complex neuronal dynamics. Interestingly, recent calculations suggest that quantum coherence can become long-lived in complex systems which are in a critical state between chaos and regularity—at the edge of quantum chaos (Vattay et al. 2014). Thus, the intricate interplay between quantum effects and nonlinear complex dynamics is able *a)* to generate new persistent quantum-chaotic patterns at a microscopic scale, *b)* to amplify quantum effects to a macroscopic scale (Satinover 2001, 209). How exactly the indeterminacy of complex quantum dynamics of the brain is embedded in classical neuronal mechanisms of decision making (Rolls 2012) remains to be determined. But we can conclude with Hermann Weyl:

> We must await the further development of science, perhaps for centuries, perhaps for thousands of years, before we can design a true and detailed picture of the interwoven texture of Matter, Life, and Soul. But the old classical determinism of Hobbes and Laplace need not oppress us any longer (1932, 65).

Acknowledgments

The author thanks Uwe Meixner for reading and improving the manuscript.

References

Aaronson, S. 2013. "The Ghost in the Quantum Turing Machine", arXiv preprint arXiv:1306.0159; to appear in *The Once and Future Thing*, a collection edited by S. Barry Cooper and A. Hodges.

Arndt, M., Juffmann, T., and Vedral, V. 2009, "Quantum physics meets biology", *HFSP Journal* 3: 386–400.

Ball, P. 2011, "The Dawn of Quantum Biology", *Nature* 474: 272–274.

Banks, T. 1998, "The State of Matrix Theory", *Nuclear Physics B-Proceedings Supplements* 62: 341–347.

Barr, S. 2003, "Retelling the Story of Science", *First Things* 131: 16–25.

Beggs, J. M., and Plenz, D. 2003, "Neuronal avalanches in neocortical circuits", *The Journal of neuroscience* 23: 11167–11177.

Beggs, J. M., and Plenz, D. 2004, "Neuronal avalanches are diverse and precise activity patterns that are stable for many hours in cortical slice cultures", *The Journal of Neuroscience* 24: 5216–5229.

Beggs, J. M., and Timme, N. 2012, "Being critical of criticality in the brain", *Frontiers in Physiology* 3: 163.

Brüntrup, G. 2008, *Das Leib-Seele Problem. Eine Einführung*, Stuttgart: Kohlhammer Verlag.

Chialvo, D.R. 2010, "Emergent complex neural dynamics", *Nature Physics* 6: 744–750.

Clarke, P.G.H. 2014, "Neuroscience, quantum indeterminism and the Cartesian soul", *Brain and Cognition* 84: 109–117.

Collini, E., Wong, C.Y., Wilk, K.E., Curmi, P.M., Brumer, P., and Scholes, G.D. 2010, "Coherently wired light-harvesting in photosynthetic marine algae at ambient temperature", *Nature* 463: 644–647.

Crick, F. 1994, *The Astonishing Hypothesis*, New York: Charles Scribner's Sons.

Davies, P. 2004, "Does quantum mechanics play a non-trivial role in life?", *BioSystems* 78: 69–79.

Deco, G., and Jirsa, V.K. 2012, "Ongoing cortical activity at rest: criticality, multistability, and ghost attractors", *The Journal of Neuroscience* 32: 3366–3375.

Dennett, D. 2003, *Freedom Evolves*, London: Penguin Books.

Doyle, B. 2011, *Free will: The Scandal in Philosophy*, Cambridge, Mass.: I-Phi Press.

Engel, G.S., Calhoun T.R., Read E.L., Ahn, T.-K., Mančal T., Cheng Y.-C., Blankenship, R.E., and Fleming, G.R. 2007, "Evidence for wavelike energy transfer through quantum coherence in photosynthetic systems", *Nature* 446: 782–786.

Frankfurt, H. 1969, "Alternate Possibilities and Moral Responsibility", *Journal of Philosophy* 66: 829–839.

Fujisawa, S., Norio, M., and Yuji, I. 2006, "Single neurons can induce phase transitions of cortical recurrent networks with multiple internal states", *Cerebral Cortex* 16: 639–654.

Glimcher, P.W. 2005, "Indeterminacy in brain and behavior", *Annual Reviews in Psychology* 56: 25–56.

Hameroff, S., and Penrose, R. 2014, "Consciousness in the universe: A review of the 'Orch OR' theory", *Physics of life reviews* 11: 39–78.

Hoefer, C. 2010, "Causal Determinism", in E.N. Zalta (ed.), *The Stanford Encyclopedia of Philosophy* (Spring 2010 Edition): http://plato.stanford.edu/archives/spr2010/entries/determinism-causal, accessed March 22, 2014.

Huelga, S.F., and Plenio, M B. 2013, "Vibrations, quanta and biology", *Contemporary Physics* 54: 181–207.

Hunter, P. 2006, "A quantum leap in biology", *EMBO reports* 7: 971–974.
Jäger, C. 2006, "Drei Konsequenzargumente für eine inkompatibilistische Theorie moralischer Verantwortung", *Zeitschrift für Philosophische Forschung* 60: 504–527.
Jedlicka, P. 2005, "Neuroethics, reductionism and dualism", *Trends in Cognitive Sciences*, 9: 172.
Jedlicka, P. 2009, "Quantum stochasticity and neuronal computations", *Nature Precedings*: http://dx.doi.org/10.1038/npre.2009.3702.1
Jordan, P. 1938, "Die Verstärkertheorie der Organismen in ihrem gegenwärtigen Stand", *Naturwissenschaften*, 26: 537–545.
Koch, C., and Hepp, K. 2006, "Quantum mechanics in the brain", *Nature* 440: 611–612.
Koch C. 2009, "Free will, physics, biology and the brain", in N. Murphy, G. Ellis, and T. O'Connor (eds.), *Downward causation and the neurobiology of free will*, Berlin: Springer.
Laplace, P.S. 1814, *A Philosophical Essay on Probabilities*. Reprint 1951, New York: Dover.
Lee, H., Cheng, Y.C., and Fleming, G.R. 2007, "Coherence dynamics in photosynthesis: protein protection of excitonic coherence", *Science* 316: 1462–1465.
Loewenstein, W. 2013, *Physics in Mind: A Quantum View of the Brain*, New York: Basic Books.
London, M., Roth, A., Beeren, L., Häusser, M., and Latham, P.E. 2010, "Sensitivity to perturbations in vivo implies high noise and suggests rate coding in cortex", *Nature* 466: 123–127.
McKenna, M. 2009, "Compatibilism", in E.N. Zalta (ed.), *The Stanford Encyclopedia of Philosophy* (Winter 2009 Edition), http://plato.stanford.edu/archives/win2009/entries/compatibilism; accessed March 22, 2014.
Monod, J. 1974, "BBC lecture", quoted in J. Lewis (ed.), *Beyond Chance and Necessity*, London: Garnstone Press.
Noble, D. 2006, *The Music of Life: Biology beyond the Genome*, Oxford: Oxford University Press.
Panitchayangkoon, G., Hayes, D. Franksted K.A., Caram, J.R., Harela, E., Wen, J., Blankenship, R.E., and Engel, G.S. 2010, "Long-lived quantum coherence in photosynthetic complexes at physiological temperature", *Proceedings of the National Academy of Sciences* 107: 12766–12770.
Penrose, R. 1996, "Consciousness involves noncomputable ingredients", in J. Brockman (ed.), *The Third Culture: Beyond the Scientific Revolution,* New York: Simon & Schuster.
Plenio M.B., and Huelga, S.F. 2008, "Dephasing-assisted transport: quantum networks and biomolecules", *New Journal of Physics* 10: 113019.
Putnam, H. 2005, "A Philosopher Looks at Quantum Mechanics (Again)", *The British Journal for the Philosophy of Science* 56: 615–634.
Rolls, E.T. 2012, "Willed action, free will, and the stochastic neurodynamics of decision-making", *Frontiers in Integrative Neuroscience* 6: 68.
Sarovar, M., Ishizaki, A. Fleming, G.R., and Whaley, K.B. 2010, "Quantum entanglement in photosynthetic light-harvesting complexes", *Nature Physics* 6: 462–467.
Satinover, J. 2001, *The Quantum Brain: The Search for Freedom and the Next Generation of Man*, New York: John Wiley & Sons.
Sension, R.J. 2007, "Biophysics: Quantum path to photosynthesis", *Nature* 446: 740–741.
Singer, W. 2009, "The Brain, a Complex Self-organizing System", *European Review* 17: 321–329.
Sompolinsky, H. 2005, "A scientific perspective on human choice", in Y. Berger and D. Shatz (eds.), *Judaism, Science, and Moral Responsibility*, The Orthodox Forum Series, Lanham: Rowman & Littlefield Publishers.
Tegmark, T. 2000, "Importance of Quantum Decoherence in Brain Processes", *Physical Review E* 61: 4194–4206.

Tiersch, M., Popescu, S., and Briegel, H.J. 2012, "A critical view on transport and entanglement in models of photosynthesis", *Philosophical Transactions of the Royal Society A: Mathematical, Physical and Engineering Sciences* 370: 3771–3786.
Van Inwagen, P. 1983, *An Essay on Free Will*, Oxford: Clarendon Press.
Van Regenmortel, M.H. 2006, "Reductionism and complexity in molecular biology", *EMBO reports* 5: 1016–1020.
Vattay, G., Kauffman, S., and Niiranen, S. 2014, "Quantum biology on the edge of quantum chaos", *PLoS ONE* 9: e89017.
Vaziri, A., Plenio, M. 2010, "Quantum coherence in ion channels: resonances, transport and verification", *New Journal of Physics* 12: 085001.
Vedral, V. 2011, "Living in a quantum world", *Scientific American* 304: 38–43.
Vihvelin, K. 2011, "Arguments for Incompatibilism", in E.N. Zalta (ed.), *The Stanford Encyclopedia of Philosophy* (Spring 2011 Edition), http://plato.stanford.edu/archives/spr2011/entries/incompatibilism-arguments; accessed March 22, 2014.
Wang, Q., Schoenlein, R.W., Peteanu, L.A., Mathies, R.A., and Shank, C.V. 1994, "Vibrationally coherent photochemistry in the femtosecond primary event of vision", *Science* 266: 422–424.
Wegner, D.M. 2003, "The mind's best trick: how we experience conscious will", *Trends in Cognitive Sciences* 7: 65–69.
Weyl, H. 1932, "The Open World: Three Lectures on the Metaphysical Implications of Science", in P. Pesic (ed.) 2009, *Mind and Nature: Selected Writings on Philosophy, Mathematics, and Physics*, Princeton: Princeton University Press.

Name Index

Aaronson, S. 188n, 190, 195-197
Adrian, E., 111
Albert, D.Z. 83
Amassian, A. 122
Anscombe, G.E.M. 174, 183
Arieli, A. 126
Aristotle 183
Aspect, A. 122
Atmanspacher, H. 63, 65, 73, 83, 171, 183
Austin, J.L. 183
Awschalom, D.D. 126
Balaguer, M. 183
Bandyopadhyay, A. 127
Barrett, J.A. 67, 76n, 77n, 79, 79n, 80n, 82n, 83
Barrow, J.D 101
Bauer, E. 61, 65
Beggs, J. 195, 197n
Bem, D.J. 124
Bennett, C.H. 122
Bianchi, M. 128
Bohm, D. 47-8, 51, 77n, 83, 132, 148n, 163
Bohr, N. 9, 13, 18, 103, 114
Born, M. 68, 72-3, 83
Brentano, F. 40
Broad, C.D. 56-7, 65
Buridan, J. 31
Byrne, P. 79n, 80n, 82n, 83
Carnap, R. 148, 151n, 163
Carter, B. 101
Chalmers, D. 41, 53-55, 57, 60, 65, 102, 132
Churchland, P.S. 124
Conway, J.H. 147, 163
Corradini, A. 2, 35, 53, 131, 159, 163-164
Craddock, T. 110, 113, 114
Crick, F. 132-133, 185, 197
Davies, P. 191, 197
Deecke, L. 121
Dennett, D. 120, 122, 124, 132, 143, 143n, 163, 187, 191, 197
Descartes, R. 67
Dirac, P. 99
Eccles, J. 139n, 164, 171n, 183

Einstein, A. 15, 18, 72-73, 76, 76n, 83, 99, 121, 126, 132, 134, 153, 190
Ellis, G.F.R. 139n, 163, 198
Emerson, D. 129
Engel, G.S. 126
Enz, C.P. 73n, 83
Epperson M. 45-46, 51
Everett III, H. 48, 51, 83, 103, 114, 133
Feigl, H. 159, 159n, 163
Frankfurt, H. 167, 169, 177, 183, 187, 188, 197
Franks, N.P. 128
Fröhlich, H. 111, 126
Gale, R. 158n, 163
Ghirardi, G. 44, 51, 83
Glimcher, P. 171n, 183, 193, 197
Goldstein, S. 40, 51, 132
Green, J.P. 128
Grünbaum, A. 161n, 163
Hagan, S. 118, 119
Hameroff, S. 46-47, 51, 101, 104, 111, 114, 116, 119, 121, 127, 129, 132-5, 171, 194, 197
Hawking, S. 16, 18, 139n, 163
Healey, R. 83
Heisenberg, M. 138n, 163, 171, 183
Heisenberg, W. 11, 141n, 163, 171n
Hiley, B. 47, 51
Hodgkin, A. 102, 107, 114, 116, 119, 125
Holt, J. 51
Houdini, H. 95
Huxley, A. 102, 107, 114, 116, 119, 125
Huxley, T. H. 120
Jackson, F. 57, 65
Jäger, C. 187-8, 198
James, W. 18, 35
Jarvik, L. 114
Jordan, P. 195, 198
Kane, R. 148n, 149, 163, 165, 176-177, 183
Kang, S. 128
Kant, I. 137, 138n, 141, 149n, 154n, 155, 163, 166
Kent, A. 83
Kim, J. 59, 65, 160, 163
Kinsbourne, M. 120, 122, 124
Koch, C. 102, 132-133, 135, 191-192, 194, 198

Kochen, S. 93, 99, 147, 163
Kornhuber, H.H. 121
Kurzweil, R. 102
Kutschera, F. von 62-63, 65
Lamme, V.A.F. 105
Lane, R.D. 129
Laplace, P.S. 146, 146n, 150n, 163, 186, 190, 196, 198
Lee, U. 105,
Legon, W. 129
Leibniz, G.W. 104
Levine, J. 57, 65
Libet, B. 122, 123, 124
Lieb, W.R., 128
Locke, J. 167, 169
Loewenstein, W. 185, 193, 198
Loewer, B. 48, 51, 76, 77n, 83
London, F. 61, 65
Longo, G. 138n, 163
Lusanna, L. 146n, 154n, 163,
Ma, Q. 125
Matsuyama, S. 114
Maxwell, J. C. 102,
Mc Fadden, J. 126
McTaggart, J. 162-163
Meixner, U. 36, 163-164, 196
Mele, A. 181, 183
Merril, C.R. 128
Misra, B. 12
Mlodinow, L. 16, 18
Molnar, G. 59, 65
Monod, J. 185, 198
Montévil, M. 138n, 163
Mossbridge, J. 124
Nagel, E. 146n, 164
Nagel, T. 54, 56, 58, 60, 65, 148, 164
Naundorf, B. 108, 125
Neumann, J. von 8, 11, 18, 61, 65, 67, 83, 103, 114
Newton, I. 9, 102, 150n
Ney, A. 77n, 83
Ouyang, M. 126
Pauli, W. 63, 68, 72-73, 73n, 83
Pauri, M. 137, 139n, 140n, 142n, 145n, 146n, 152, 154n, 161n, 163-4
Peirce, C.S. 28

Penrose, R. 46-47, 51, 102-4, 114-116, 119-120, 122, 127, 133-34, 171, 183, 191, 194, 197-198
Peres, A. 125
Planck, M. 46, 103, 115
Plenio, M. 193, 197-199
Plenz, D. 195, 197
Plumhof, J.D. 126
Pockett, S. 126
Podolsky, B. 83, 121, 132
Pollen, D.A. 122
Popper, K.R. 8, 139, 164
Primas, H. 73, 83
Putnam, H. 153, 153n, 164, 190, 198
Rasmussen, S. 111
Ray, P.G. 122
Rimini, A. 44, 51, 83
Roelfsema, P.R. 105
Rosen, N. 76n, 83, 99, 121, 132
Ryle, G. 167
Satinover, J. 189, 192, 194, 196, 198
Sahu, S. 119, 127
Saunders, S. 79n, 80n, 83n
Schrödinger, E. 32, 38, 44, 103, 116, 120
Seager W. 54, 65
Sellars, W. 159, 163, 165
Shimony, A. 104, 141n, 143n, 153n, 164
Siewert, C. 42, 51
Singer, W. 187, 198
Smith, S.A. 111
Snyder, S.H. 128
Sompolinsky, H. 193-194, 198
Sri Aurobindo 92, 94, 96-97, 99
Stapp, H.P. 7, 16, 18, 32, 32n, 33, 33n, 34, 34n, 36, 44-45, 51, 61, 103, 114, 135, 171
Strawson, G. 54-55, 57-58, 65
Strawson, P.F. 165, 183
Sudarshan, G. 12
Tegmark, M. 118, 192, 198
Tipler, F.J. 101
Tittel, W. 122
Tononi, G. 107
Torretti. R. 145, 155n, 156n, 164
Troutt, L.L. 128
Tyndall, J. 7, 18
Van Inwagen, P. 187, 199
Watt, C. 111

Vedral, V. 193, 197, 199
Velmans, M. 120, 122
Wallace, D. 19, 83
Weber, T. 44, 51, 83
Wegner, D. 120
Weinberg, S. 143, 143n, 160, 160n, 161, 164
Weyl, H. 190, 196, 199
Whitehead, A.N. 18, 45, 51, 104, 135, 140n,
    145, 147, 148n, 158n, 164
Wiesner, S.J. 122
Wigner, E.P. 67, 69-72, 78, 81, 83, 103, 114,
    148n, 164
Wittgenstein, L. 167
Yaron-Jakoubovitch, A. 126
Zeh, H.-Z. 48, 51
Zeno of Elea 12, 33-35

Subject Index

Action, 138
Actors and spectators, 13
Actuality, 46
Actualization(s), 12, 15, 139
Alternative possibilities, 21, 26, 32-33
- requirement of (AP), 168-170
ānanda, 98
Animal, 19, 27, 31
Antecedents, 149
Appearances, 143
- are deceiving, 14-15
Arrow in flight, 12
Astrophysics, 145
Atomization, 138
Attributes, of events, 152

Basic realities, 9
Becoming, 158
Being, 87, 91, 92, 94, 97, 98,
Biological evolution, 31
Bivalence principle, 162
Bohmian mechanics, 77-79, 81-82, 190
Brain(s), 11-15, 17, 19, 25-27, 31-32, 94, 95, 137
Buridan's ass, 31

Cause, 19, 30-31, 34-35
- efficient, 144
- final, 144n, 149, 151
Causal chain, 148, 155
- emergent, 156, 158
- physical, 149, 155, 157
Causal,
- closure, 34-35, 150
- closure, principle of, 37
- effectiveness of mental intent, 12-13
- structure, 153
Causality,
- through freedom, 138
- physical, 140, 141, 150
- absence of, 150
Causation,
- agent, 148
- non occurrent, 148
- downward, 149

Chance, 29-35, 35n, 137
Chaos, chaotic, 145n, 194-196
chit, 98
Choice, 32-34, 33-34n
Classical,
- mechanics, 7, 8, 9, 17
- physics, 1, 87
Collapse,
- of the quantum state, 10, 14, 16
- of Schrödinger's wave function, 39-40, 44-47, 48, 49, 189, 190n
Combination problem, 60, 104
Compatibilism, 142, 165-166, 187-189
Complexity, 58-59
- theory of, 195-196
Conditionals, 20-21, 24, 26-28
Conjuring trick, 87
Conscious,
- soul, 35
- thoughts, 1, 8
Consciousness, 19, 29, 31-32, 35, 37, 40-50, 53, 61, 65, 91-94, 97-98, 102-106, 108, 112, 119-125, 138
- evolution of, 94-95
- exclusive concentration of, 92-93, 94, 95, 98
- mental, 92, 95
- multiple concentration of, 92
- subliminal, 93, 96, 97
- supramental, 92, 94
- surface, 95, 96, 97
- transformed, 98
Consequence argument, 187-189
Conventionality of distant simultaneity, 153
Copenhagen quantum mechanics, 8, 11, 14
Cosmic time, 158
Cosmology, 145, 146
Critical/subcritical/ supracritical, 195-196
Criticality, 195

Decision making, 96
Definite states, 29
Detection of indetermination, 19, 24-30
Detector, 89

Subject Index

Determinism, 21, 35, 72, 137n, 140-141, 165-172, 185, 185n, 186-196
Deterministic grand reductionism, 95-96, 143, 160
Directedness, 59
Dispositions, 60
Distinctions,
- between regions of space, 88-89
- between things, 89-90, 92

Domain of validity, 146n
DOMINDARs, 19-20, 24-27, 27n, 28-30, 30n, 31-35
Dual-aspect theories, 63
Dualism, 19
- property, 53-54
- substance, 53-54, 56, 65
- polar, 62-63
- psycho-physical, 137-138, 159
- physical-nonphysical, 68, 73, 73n, 74 76-77, 80-81
- physical-physical, 77-78, 80-82

Effort of will, 172-180
Ego, 11, 12
Electro-encephalography (EEG), 106, 118-119, 126-127
Emergence,
- brute, 57-59
Emergentism, 53, 56-59
Empirical consistency, 74
Entanglement, 191, 193
Epiphenomenalism, 53, 65
Event, 149, 152
Evolution, 92, 94-96, 98
Executive force, 92
Experience, 7-10, 14
- structure of, 9
Explanation, 150
Explanatory gap, 159
Expressive ideas, 92
External (living) superobserver, 154, 156

Finality, 139
First person point of view, 56
Free,
- choices, 9, 11, 12, 15
- will, 95, 96, 119-121, 125, 130, 137, 138n, 139, 172-182, 185, 187-190, 194-195

Freedom, 29, 33, 92, 95, 96, 138
- and action, 166-170
- and chance, 179-181
- and control, 179-180
- and luck, 179-182
- and neuroscience, 170-172
- and will-setting, 177-179
- genuine, 95
- illusionary, 96
- infinite, 96
- libertarian, 165-182
- of action, 139
- of choice, 139
- of meaningfulness, 139

Fundamentality of the mental, 53-54, 60, 62

Galilean blockade of essential individuation, 140, 144
General relativity, 93, 146n
Global instant "Now", 17
Grain objection, 143n
GRW, 78, 81

Hamiltonian system, 145
Heisenberg cut, 11
Hidden variable, 37, 39, 47-48
History created by us, 16
Holism, 137
Human action, 32-34

Illusion (universal)/ illusions, 14-15, 140, 143n
Incompatibilism, 142, 166-170
Inconsistency, 11
Indeterminism, 21, 94, 96, 137, 140-141, 165-182, 189n, 190
Indian philosophy, 93, 95
Individuals, 92-93, 94
Individuating,
- properties, 90
- substances, 90
Inference,
- of indetermination, 23, 26-30
- to the best explanation, 19
Infinite
- delight, 92, 97

- force, 93
- quality, 98, 100, 104
Information, 40-45, 47-48, 149, 149n
Initial conditions, 139, 145, 147, 147n, 156
Input-output, 26, 28
Inscrutability, of matter, 159
Intention, 32-33
Intentionality, 40-50, 97
Interference, 38, 39
Interpretative principle, 88, 89
Intrinsic nonlocal character of nature, 15-16
Intrinsicality, 59
Introspection, 159
Involution, 92, 98, 100
Ion channels, 15
Jung-Pauli theory, 63

Kantian oversight, 154
Knowledge,
- by identity, 97
- direct, 97

Libertarian, 172-182
Libertarianism. See free will, libertarian.
Liberum arbitrium voluntatis, 181-182
Life, 94
Light-cone, 153
Lila, 93, 95
Literal (aporetic) interpretation of reduced ontology, 142, 142n, 153
Living beings, 138, 142, 147
Local physics, 145
Logical circle, 91
Luck objections (to free will), 181-182
Lyapunov exponents, 145n

Macro-objectification, 138n
Macroscopic
- and microscopic indetermination, 19, 23, 24, 26, 27, 30
- and microscopic object, 89
- position, 89
Macroworld, 91
Magnetoreception, 193
Man,
- as object, 7, 10, 11
- as subject, 7, 10, 11

Manifestation, 90-92, 94, 95, 96, 98
Many-minds theory, 77n
Many-worlds interpretation, 77, 79, 189-190, 190n
Material composition, 54-55
Materialism, 1, 35, 54-56, 62, 65
Meaningless life, 10, 17
Measurement, 38-39, 48
- outcomes, 88-93
- pointer, 87
- problem, 87
Measuring devices, 11, 14
Mental,
- control of bodily motion, 12, 17
- intentions, 9, 11, 13, 14
Metaphysics vs. meta-physics, 140, 149, 149n, 150-152, 154, 159
Microtubules, 104, 107, 109-114, 116-119, 126-131
Mind-body,
- co-evolution, 59
- dualism, 67-68, 72, 75, 78, 81
- interactionism, 53, 62
-nexus, 1
Mind, 94
Minimality (natural),
Minkowski spacetime, 152, 154
Monism, 53, 65
- constitutive Russellian, 53
Moral responsibility, 187-189
Mythological explanations, 98

Natural souls, 19
Nature, 19, 32-34
Nature's response, 12
Neural structure, 138, 138n
Neurodeterminism, 186-187, 188-189, 190
Neuronal avalanches, 195
Neurophilosophy, 1
Neuroscience, 1
Neutral monism, 63-64
No-branching, 74, 79
No-collapse, 68, 72-73, 75-76, 78-81
No-go theorems, 93
Non-deducibility, 57
Non-emergence, 54, 57
Non-explicability, 57

Nonlinear dynamics, 192, 194-195, 196
Non-locality, 15-16, 39
Non-predictability, 57
Nonreductionism, 54-56
Non reductive philosophy of mind, 1
Nontrivial quantum effects, 191-192, 193
Novelty, 57
Nowness, 143n, 150

Objectification, 140
Observer, 145
- living, 153, 153n, 154
- fictional, 153, 153n, 156
Observer's values, 9, 10
Olfaction, 193
Ontologically vague states, 26n
Ontological-phenomenological aporia, 140
Open world, 140, 144, 147, 154, 158
Orch OR, 104, 114-120, 126-127, 131
Organism, 25, 31, 139, 158
Orthodox quantum mechanics, 15, 17

Pairing problem, 59
Panprotopsychism, 65
Panpsychism, 43, 46, 47, 53-62, 65
Parallel processing, 175-178
Parsimony, 15
Participants, 9
Particles,
- elementary, 93
- formless, 91, 93, 94, 98
- fundamental, 90, 91, 92
- numerical identity of, 90
Peircean abduction, 28
Perceptions, 8-9, 11, 15
Permanent actuality, of events, 153n
Phenomena, 7, 9, 16
Phenomenal properties, 159
Phenomenological
- arrow of time, 142
- time, 142
Photosynthesis, 192
Physical,
- arrow of time, 142
- description of the world, 140, 147
- laws, 139, 145
- object, 89

- pluralism, 80
- time, 145, 152
- world, 91, 97
Physicalism, 1, 9, 13
- broad, 55
- narrow, 55
- nonreductive, 53
Pilot wave, 40, 47
Plural voluntary control (PVC), 176, 180
Postulate of freedom, 140-141, 144, 145n, 153
Potentiality/potentialities, 15, 16, 17, 46, 139, 141n, 159
Powers, 59, 65
Practical tool, 9, 10, 11
Pragmatic,
- approach, 9
- balance, 154
Prakriti, 95, 96
Predetermination, 22, 24-25
Pre-scientific era, 98
Privacy,
- principle of, 60
Probability, 29, 34
Probing action, 9, 10, 32-34
Promissory materialism, 7, 13, 16
Property attribution, 89, 91
Pure wave mechanics, 79, 82, 82n
Purusha, 95, 96
Pyramidal neurons, 106-108, 119, 126

Qualia eliminativism, 143n
Qualities,
- primary, 60-61
- secondary, 60-61, 64
Quantum
- amplification, 195
- biology, 193
- brain, 190-191, 194-195
- chaos, 196
- coherence, 192-194, 196
- computing, 116
- fluctuations, 191-192, 194-196
- measurement problem, 67-69, 70, 72, 78
- mechanics, 88, 89, 90
  - alternatives in, 88, 89, 90
  - amplitudes in, 88, 90
  - collapse of states in, 97

- correlations in, 94
- explanatory arrow in, 87, 91, 98
- indeterminism in, 98
- irreducible empirical core of, 87
- mathematical formalism of, 87
- standard von Neumann-Dirac formulation of, 68-69
- neurobiology, 192-195
- physics, 8, 13, 14, 165-166, 172-173
  - subjectivist interpretation of, 63
- theory, 146
- uncertainty, 15
- Zeno effect, 12
Quiddities, 55

Rational(ity), 19, 24, 31, 35
- of free choices, 178-179
REACTORs, 20, 20n, 21, 21n, 22, 22n, 23-25
Real time, 158
- footprints, 142
Realism, 54, 56, 63, 72
Realistic formulation of von Neumann's Quantum Mechanics, 9
Reality,
- measurement-independent, 87
- mind-independent, 87
- psycho-physical, 11
Records of the past,
- inaccurate, 17
Reduced ontology, 140, 154,
Reductionism, 137, 185n
Representations, 97
Restriction of indetermination, 19-20, 23-24, 29-31
Retro-causal phenomena, 16-17
Rhodopsin, 193

sat, 98
Schrödinger's
- cat, 21-22, 26
- wave function, 190
Science, 7-9, 13, 14, 17
Selection-actualization, 26
Self-determination, 148
Self-formation, 169, 173, 177
Self-forming actions (SFAs), 172-182
Self,

- conscious, 56
- phenomenal, 61
- actualization of, 59
- particularization of, 59
- individualization of, 59
Sentience, 139
Single-mind theory, 77n, 78, 81
Soul, 55-56, 65
Space,
- parts of, 89
- regions of, 88, 89
Spatial differentiation,
- incompleteness of, 89, 90, 91, 97
Spatial relations, 91, 93, 94, 98
Spatialization of time, 153
Spatiotemporal homogeneity, 145
Special relativity, 141-142, 143, 152-154
Spectators, 9, 13
Standard model, 93
State completeness, 74, 76-77, 82
Statistical explanation, 33
Stochastic, 185-186, 194
Structure of experience, 9
Subject-object dichotomy, 92
Subject(s),
- of consciousness, 19, 29, 32
- of experience, 56, 60
Subjectivity, 139
Substance,
- material, 57
- mental, 56
Sufficient reason,
- principle of, 37
Superposition, 26-27, 29-30, 39, 44, 46, 48
- principle, 138
Supervenience, 43
- of the mental on the physical, 53

Temporal anisotropy, 158
Temporal determinations,
- tensed, 151, 158, 162
- tenseless, 142, 151, 162
Temporal repeatability, 145
The set W of all real events of the world, 152, 162
Third antinomy of reason, 149, 154, 156n

Trans-physical vs. intra-physical emergent
    laws, 56, 58
Transiency, 150
Two-slit experiment, 88

Ultimate,
- originator, 144
- reality, 92, 98
- responsibility (UR), 168-169
Unanswerable questions, 90
Uncertainty principle, 89
Undetermination, 147
Unpredictability, 155, 157, 158

Value-based efforts, 10
vedanta, 98
View from everywhere, 92, 94

Wave, 10
- package, 39
- function, 40, 44, 47-49
Wigner's friend, 69-70, 74, 77
Will, 166-182
World-line, 152n

The Contributors

Jeff Barrett, University of California at Irvine, USA

Godehard Brüntrup, School of Philosophy, Munich, Germany

Antonella Corradini, Catholic University of Milan, Italy

Stuart Hameroff, University of Arizona, Tucson, USA

Peter Jedlicka, Goethe-University, Frankfurt, Germany

Robert Kane, University of Texas at Austin, USA

Uwe Meixner, University of Augsburg, Germany

Ulrich Morhoff, Sri Aurobindo Ashram, Pondicherry, India

Massimo Pauri, University of Parma, Italy

Henry P. Stapp, University of California at Berkeley, USA

www.ingramcontent.com/pod-product-compliance
Lightning Source LLC
Chambersburg PA
CBHW050902160426
43194CB00011B/2255